Contents

Acknowledgements

Introduction

1. Article 987. Trovants: rocks that grow: core matter, water and human consciousness
2. Article 988. Trovants, Planet X and how planets continuously grow in size
3. Article 1005. Planet X in the sky and California earthquakes
4. Article 1012. Planet X causes spherical object to emerge from the ground
5. Article 1065. Planet X causes volcano formation by pulling cores to the surface
6. Article 1082. Planet X effects on Mars: volcanoes and craters
7. Article 1134. Intelligent life all over the Universe
8. Article 1140. Spherical objects and soil emerging from ground in California due to Planet X
9. Article 1144. The Universe is alive: God is reproducing Himself
10. Article 1147. Planet X effects accelerating: water emerges in California
11. Article 1150. Earthquakes: Planet X effect on earth's cores: not tectonic plate movement
12. Article 1154. Major Planet X activity in California due to Planet X
13. Article 1157. Small sinkholes at cemetery in California are not gopher holes
14. Article 1172. Earth cores emerge from the ground in California due to Planet X
15. Article 1176. Earth cores emerging and sinkholes forming in California due to Planet X
16. Article 1181. Planet X depleting California cores: darkness and light sinkholes
17. Article 1185. Planet X activity in California and the amazing universe we live in
18. Article 1193. Roll clouds: Planet X core systems broke apart
19. Article 1195. Sinkholes: caused by Planet X and show that the universe is alive
20. Article 1168. Planetary wide Planet X cataclysm starting in California
21. Article 1141. Asteroids coming in due to Planet X: Earth being destroyed
22. Article 1192. Fires in California: magma rain due to Planet X in the sky
23. Article 1199. Alarming surface reformation occurring in California due to Planet X and life

Books previously published

Book 1: Planet X: the awakening is now.
Book 2: The Planet X Report 2017: Photographic Evidence.
Book 3: Planet X Revealed Gravity and Light.
Book 4: The Sun Simulator
Book 5: Chemtrails: The Silent Killer.
Book 6: Planet X Physicist Articles: Part 1
Book 7: Planet X: The effects on the Earth and the Sun
Book 8: Planet X and the Solar System
Book 9: Planet X and the Hurricane Michael Cover-Up
Book 10: Planet X Reveals How the Universe Works
Book 11: Planet X and the Bible
Book 12: Planet X: the Greatest Cover Up in World History

Acknowledgements

This book would not be possible without the photographs that Beth sent me and also the effort and time that she took to investigate what was occurring in her property and surrounding area. So a big thank you to you Beth for taking this journey of discovery with me.

Introduction

This book is made up of the articles I have written describing what has been happening on a property in California. Beth, who lives on this property, is an extremely observant lady and she has been sending me photographs of both the skies, over her location, and the amazing activity that has been occurring on the ground, as spherical objects and newly created soil continuously emerges from the ground in addition to new sinkholes appearing daily.

The studying of these photographs has helped me understand what is really occurring on this planet, as well as understand the real laws of the universe. Planet X is the core systems or remains of the destroyed Planet X planets. They were destroyed as a result of energy depletion and are coming to the earth to absorb energy. Planet X's effect on the earth are not benign, the ultimate result of the energy absorption process is the destruction of planet earth just as the Planet X planets were destroyed, but at the same time, the process is revealing the truth about how the universe really operates. The observations emerging from this property in California lead to the conclusion that the earth as a planet is alive, it is a living organism. And if the earth as a celestial object is alive so are stars and galaxies as they all exist as a result of having core systems inside them just like the earth. In addition, star core systems eject core matter which turns into planets just like galactic cores eject matter which turns into globular clusters or groups of stars and so the whole universe is connected and is therefore one extremely large living organism.

Dr. Claudia Albers

Planet X Physicist

October 14th 2019

Chapter 1

987. Trovants: rocks that grow: core matter, water and human consciousness

Trovants are a particular type of rock, which seems to mainly occur in Romania but has also been reported in other parts of the world. These rocks grow, move and multiply. The rocks have a sandstone nucleus surrounded by a sand layer and when exposed to water create smaller versions of themselves, which eventually fall off and become a new trovant. The rocks also secrete sand when exposed to water.

Figure 1.1. The rocks are cylindrical, spherical or ellipsoidal in shape. The small dots on the surface are new secretions of sandy material.

Figure 1.2. The inside of a trovant: the inner structure seems to conform to a cylindrical symmetry: the rock is made of circular sheets of rock with the central sheets having a larger radius than the outer ones which would make the overall shape toroidal.

Figure 1.3. Trovant babies grow out of a large trovant. Once they reach a certain size the small trovants seem to break off the parent and become independent trovants.

Now, normal rocks do not grow and so these are not normal rocks. These rocks create matter. Normal rocks get wet all over the world without ever growing so these rocks are creating matter, they are transforming water into rock and the rock they create is the same type of rock that they are made of. They are therefore creator matter or core matter (see Article 953: Planet X: core matter and the basic electrogravitic drive) [1].

Now, Planet X observations have led to the discovery that all celestial objects have cores, which create matter, such as magma, water, oil and sand. Cores are made of rocky solid matter. The moon is a central core and it is rocky so it is likely that all cores are rocky. Cores seem to create matter by converting energy (photon energy or light) within the material, which they are made of into matter. Magma emerging from inside the earth is in a high energy state, as it is very hot and it is essentially molten rock which is created by a magma creating core, within the body of the earth (see Article 785: Planet X is here but what is it exactly?) [2].

Figure 1.4. A Planet X core in the earth's atmosphere: it is a spherical solid object surrounded by a cloud envelope, which will most likely be made of water vapor suggesting that this is a water creating core but most of the cores seem to create water as well as other materials.

The fact that the trovants create sand and that they are essentially made of sand suggests that they are sand creating cores and that cores are likely to be made of the same type of material, which they create. So, magma creating cores would be made of granite, water creating may be made of a form of rock which contains water, perhaps a type of crystal. But the trovants seem to only grow by a significant amount when they are exposed to water, which suggests that they convert water to sand, a transformation, which undoubtedly would take energy indicating that there is energy inside these rocks, but also shows that water is a very special kind of material, which can be transformed by core matter into other materials. It suggests that planets may actually be made of water, water that is transformed into rock. This would then explain why in Genesis verse 2, we read that the spirit of God was upon the face of the waters, when God was about to recreate the earth, after it had been destroyed with the other Planet X planets.

1 In the beginning God created the heaven and the earth. 2 And the earth was without form, and void; and darkness was upon the face of the deep. And the Spirit of God moved upon the face of the waters.

3 And God said, Let there be light: and there was light. (Gen. 1:1-3)

God first created the universe, and then we read that the earth was without form or destroyed and in chaos, which could not have been its initial state. The timespan between these two statements could have been billions of years. Then the Spirit of God was moving upon the face of the waters. We know that a destroyed planet becomes a chaotic jumble of cores and debris all enveloped in cloud, which is mostly water vapor, but this may actually be referring to everything being made of water. The light is the energy, which entered the earth's core system and caused the planet to come together again.

But the whole universe, as I have shown in previous articles is made out of light, as all energy and matter is essentially made from light (see Article 818: Planet X observations: the electrical universe is made of light) [3], which would then suggest that water is an intermediate material, between free photons and rocky material. It suggests that when light becomes matter it first turns into water, and then from water, into all kinds of other rocky materials. This would then explain why water responds to frequencies, adopting different crystal structures, when it then freezes. In addition, Masaru Emoto, a Japanese scientist, found that water responded to human consciousness, so that a different structure could be evoked in ice crystals, by placing a label with a particular word on a bottle of water. This then means that since all rocky material and essentially all that exists, on this planet, seems to be made of water, everything will respond to human consciousness, or to words, thoughts and feelings, flowing from human beings. Words of love and kindness produce life and complex structure, whilst words of hatred create chaos. This would agree with human beings being created in the Creator's image and the earth being a training ground for Creator children (see Article 957: Earth: training ground for Creator children not aliens) [4].

Figure 1.5. Crystal structure of water when exposed to different sounds and thoughts, either through speaking words, near the water, or by placing a written label on the water. So, it is not even the sound of the words but the sentiment and intention flowing from a human being responsible for speaking or writing the word or perhaps creating the music.

In conclusion, rocks called trovants, which move, grow and multiply seem to be core matter, or sand creating cores, and they seem to transform water into sand. The universe is made of light, and water seems to be an intermediate material out of which all core matter and created matter is made of. Thus, when light becomes matter, it seems to first turn into water and then into other materials. This would then suggest that since water responds to human consciousness, all matter on this planet will respond to human consciousness.

References:

[1] Albers, C. (2019). Article 953: Planet X: core matter and the basic electrogravitic drive.
[2] Albers, C. (2019). Article 785: Planet X is here but what is it exactly?
[3] Albers, C. (2019). Article 818: Planet X observations: the electrical universe is made of light.
[4] Albers, C. (2019). Article 957: Earth: training ground for Creator children not aliens.

Chapter 2

988. Trovants, Planet X and how planets continuously grow in size

Trovants are rocks that grow, multiply and move and are thus sand creating cores which somehow have ended up on the surface of the earth rather than inside the body of the earth like the rest of the earth's core system. They have a sandstone nucleus and an outer layer of sand. They thus seem to secrete or create sand and also create other cores, with sandstone nuclei, which are smaller versions of the sand creating cores. The sand is created matter and the baby trovants are creator matter (see Article 987: Trovants: rocks that grow: core matter, water and human consciousness) [1].

Figure 2.1. Sand secretions on a trovant rock: a sand creating core. The large trovant closest to the camera looks like a peanut shell or like two spheres joined together.

These sand creating cores show that cores continuously create matter and multiply, and that therefore, planets are supposed to continuously grow in size. The blobs of matter on the surface suggest how the water creating Planet X Stellar Cores, energy depleted cores from the Planet X planets, may look like as they create water: water appears as drops of water on the surface of the object. They also indicate why Planet X SCs can come in many different shapes as these cores also occur in many different shapes.

Figure 2.2. A fat disk shaped sand creating core on the surface of the earth with flatter disk shaped cores further back. Some cores are sticking out of the sand, suggesting that they have grown from other cores that are deeper inside the earth. A circular piece of sand on the closet trovant or sand creating core suggests that it is a recent secretion or an example of matter creation

Figure 2.3. A Planet X SC in the shape of a disk, in the sky, giving rise to a cloud in the shape of a disk. The cloud is the object's cloud envelope and suggests that this is primarily a water creating core as cloud seems to be made of water vapor (see Article 785: Planet X is here but what is it exactly?) [2].

Figure 2.4. Trovants seem to have a sand outer envelope suggesting that these cores create either no water or very little water but they transform water created by other cores into sand.

The fact that these cores create sand out of water, suggests that where there is a lot of sand, such as in a desert region, there is most likely a lot of sand creating cores and not many water creating cores, and these sand creating cores transform most of the water, coming from the water creating cores, into sand.

Figure 2.5. Sand on the surface of the earth, in a desert region, which must have been created by sand creating cores, inside the body of the earth, which most likely have transformed water, created by water creating cores, into sand.

The fact that sand creating cores transform water into sand suggests that water creating cores are a major part of a planetary core system, with most cores, such as magma creating cores, being able to create water as well. In addition, it is likely that the magma creating cores start off creating water and then transform some of the water they create into rock, which comes out as molten rock.

In conclusion, trovants, which seem to be sand creating cores on the surface of the earth, suggest why Planet X Stellar Cores come in different shapes and that planets continuously grow. And if planets continuously grow, so do stars and galaxies, and the whole universe.

References:

[1] Albers, C. (2019). Article 987: Trovants: rocks that grow: core matter, water and human consciousness.
[2] Albers, C. (2019). Article 785: Planet X is here but what is it exactly?

Chapter 3

1005. Planet X in the sky and California earthquakes

Figure 1 below shows a photograph sent to me by Ryan of the sky in California. It shows the typical cloud formation characteristic of the large Planet X Stellar Cores when they penetrate into the earth's atmosphere. These objects are huge and emit blue light from their surfaces but the objects' cloud envelope is sparse and typically has a clumpy aspect, which can be seen in this case, above some lower altitude clouds. In addition, the Sun appears to be a reflection from off the cloud surface, which is also typical of these objects being in the sky, as the simulators, in orbit, cannot be seen through such a huge solid object.

Figure 3.1. Photograph of the sky sent in by Ryan, over Citrus Heights California, the photograph was taken in the morning, on July 16th 2019. The top layer of cloud seems to be at one uniform height with blue seen in between, a typical cloud formation observed when a large (huge) Planet X Stellar Cores (SCs), the energy depleted cores of the destroyed Planet X planets and stars. The objects emit blue light from their surfaces, and it is thus the surface of a solid object that is seen between the clumps of cloud. In this case, the surface seems to be dark blue, which suggests that the object is still extremely depleted in energy, as they seem to get lighter blue, once they have been inside the earth's atmosphere for some time. The sun is usually simulated below the objects which makes the 'sun' look like a reflection off the surface of the object, or its cloud envelope, which is also what seems to be happening here.

Figure 3.2. The cloud pattern above is similar to this one over Hawaii from June 20th 2019. The Sun is a reflection off the surface and the clumpy clouds indicate that we are looking at the solid blue light emitting surface of a large Planet X SC (see Article 992: Huge Planet X object in the sky over Hawaii: causes dizziness and nausea) [1].

Figure 3.3. One of the first images I saw that made me realize that these objects were in the sky, at very high altitudes, but certainly inside the atmosphere, or at least a part of the objects were penetrating the atmosphere down to the level where they could be observed and the artificial sky simulation system was not able to hide them. What seems to be a blue break in the clouds is actually the surface of a solid object and it is curved downwards, the cloud spout indicates this clearing to be a connection point. The object creates this type of cloud at the gravitational connection point and the spout is reminiscent of a tornado spout but much smaller. The cloud envelope cloud elsewhere is sparse and clumpy with lots of spaces through which the blue light emitting surface is visible. The 'Sun' seems to also be a reflection off the surface of the object (see Article 786: Sky blue Planet X inside the earth's atmosphere) [2].

The lighter blue color emitted by the surface and the pink light emitted by its cloud envelope indicate that the object has been in the earth's atmosphere, for quite some time and has been able to attain a higher level of energy, than objects that emit a darker blue light from their surfaces.

Figure 3.4. Another photograph from Ryan showing the clumpy cloud envelope, which seems to be denser than with other objects, observed previously, but which can still be identified as a typical cloud envelope. The object's darker blue surface and denser cloud envelope suggests that these objects slowly lose their cloud envelopes over time, as they absorb more and more energy from the earth's core, thus explaining why there is often cloud material in the earth's atmosphere, which is not associated to any core in the sky.

Figure 3.5. Another photograph from Ryan: It clearly shows the clumpy, but denser than usual cloud envelope, and the fact that the surface is much darker blue than the clear sky around it. The cloud is also clearly at an extremely high altitude, which is also typical of these objects, as the larger they are, the stronger seems to be the repulsive force, which stops them from getting any closer to the earth's surface.

Figure 3.6. Photograph, sent in by Beth, of the sky over the Santa Rosa region, and showing a very similar cloud pattern: This photograph was also taken on July 16th 2019, in the morning. The cloud formation is also denser than usual and clumpy with a dark blue color seen between the clumps of cloud. However, the cloud seems to be at an even higher altitude in this photograph suggesting that they were taken earlier and the object was at a higher altitude then. This would then suggest that these objects slowly move down through the atmosphere to a minimum altitude.

Figure 3.7. Another photograph from Beth which clearly shows that there was a huge Planet X object in the sky in the morning of July 16th 2019 over California.

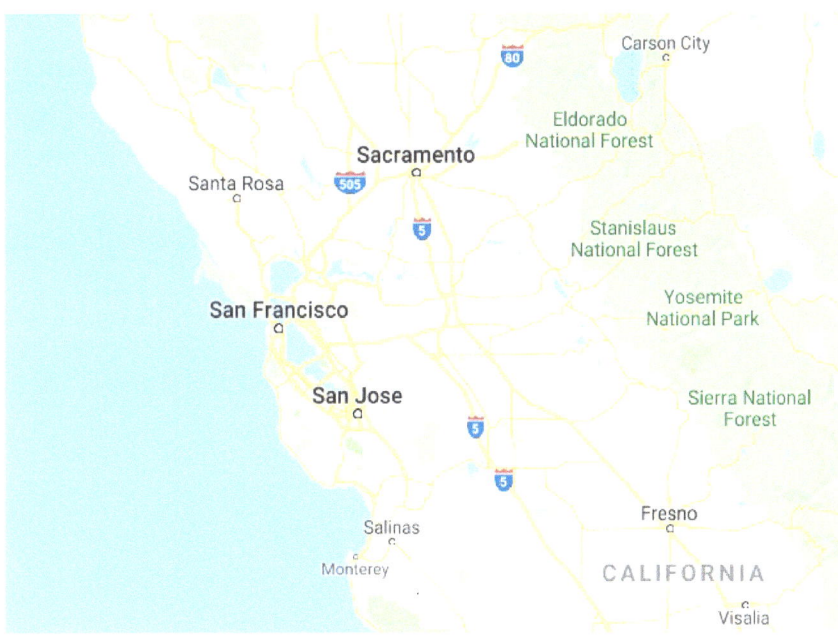

Figure 3.8. Beth's photographs were of the sky over the Santa Rosa region, which is north of San Francisco.

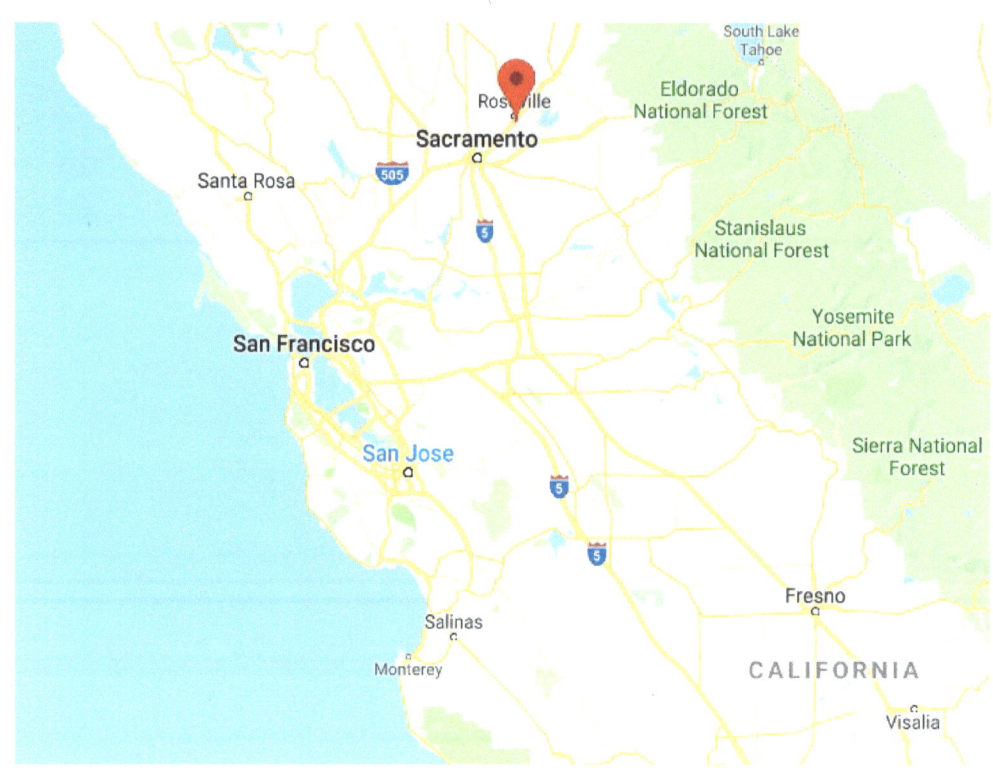

Figure 3.9. Ryan's photographs were from Citrus Heights which is northeast of San Francisco, 250 km (155 miles) away.

So, it seems that the same object was seen in the sky from 2 locations, separated by a distance of 155 miles, which indicates that it was a huge object, most likely over 1200 miles in diameter. These huge objects make gravitational connections with the earth's central core, which then induce earthquakes (see Article 1003: Sinkholes in California close to earthquake swarm region: rift unzipping) [3] And on July 16th 2019 there were two earthquakes close to San Francisco, which suggests that this particular object was responsible.

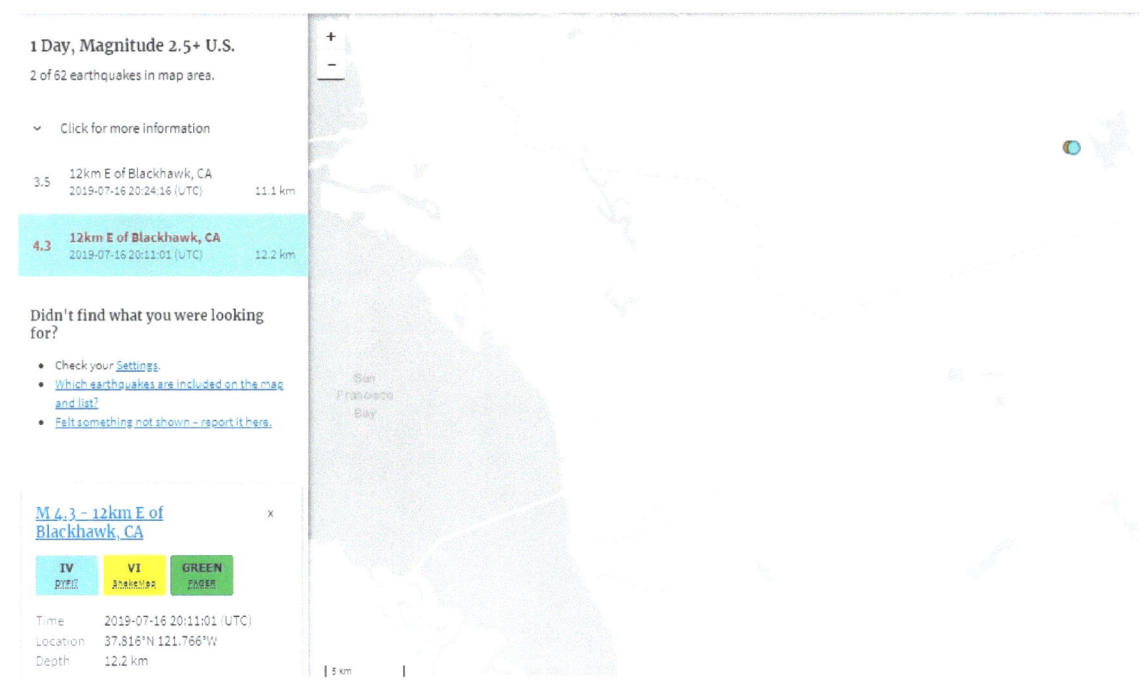

Figure 3.10. There were two earthquakes close to San Francisco which is not far from where this object was seen which suggests that it was responsible.

The Large Planet X SC seems to still be depleted in energy and it is thus likely that as it continues to absorb energy from the earth's core that the gravitational connection with the earth's central core will increase, at which time it may induce stronger earthquakes. Since the object was seen close to San Francisco and two earthquakes occurred close to San Francisco the connection point is most likely very close to San Francisco. Either an old or a new sinkhole will most likely have formed at the center of the connection point similarly to the Birch Bay connection point.

Figure 3.11. The rounded shape of Birch Bay suggests that it originally formed as a result of Planet X making a connection at this location, which led to the formation of a sinkhole. The shape of the ground under the water appears to be in the form of 2 other circular outlines indicating that another 2 smaller sinkholes appear to have formed within the larger outline of the Bay, which suggests that smaller objects have made similar connections at this location. This the most likely center of the connection which led to the 4.6 magnitude earthquake close to Seattle on July 12th 2019.

These objects through their connections with the earth's central core induce surface reformation events, which are caused by the central core ejecting core matter, which then changes the surface gravitational field and configuration and often leads to surface expansion, rifts or fissures opening and even new oceans being created. Because this region is so close to the coastline, it is likely that this surface reformation at this region will just increase the size of the Pacific Ocean so that this part of the landmass ends up a part of the ocean (see Article 997: Earthquake in Seattle on July 12th ground dropping and core reset event) [4].

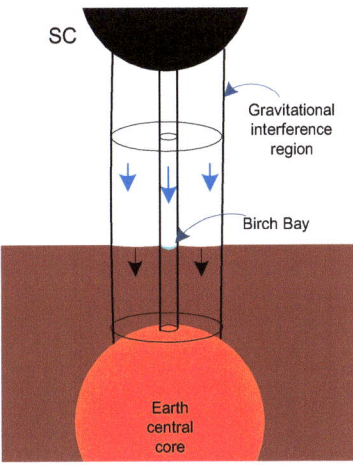

Figure 3.12. Illustration of the connection, but not to scale, as the outside ring region is some twenty times larger than the inner part. The field is still downwards in the outer ring, which would cause the ground inside it to have a tendency to move downwards, thus explaining the movement of the ground during the 4.6 magnitude earthquake (see Article 1000: Birch Bay tides connected to Seattle earthquakes and due to Planet X) [5].

Figure 3.13. The pattern of earthquakes all along the coast of California suggest that a large scale surface reformation event is starting to occur in this region of the world, which is likely to cause a rift to unzip across the whole of the California coast as well as through the Ridgecrest region. As the huge Planet X SC observed in the sky over California increases in energy it is likely that the severity of the earthquakes increases and that larger and larger fissures form and that also the ground that already has a tendency to sink sinks by greater and greater increments with each event.

This unfolding event is likely to be cataclysmic for the West Coast of the US and will most likely lead to a worldwide event in that the earth's central core goes through massive core ejection which then changes most of the surface of the earth as well as resets the core's and thus the earth's axis of rotation, i.e. a sudden Pole Shift event.

In conclusion, huge Planet X Stellar Cores were observed in the sky close to San Francisco on the same day that there were 2 earthquakes close to San Francisco suggesting that the object was the cause. This object is likely to cause more severe earthquakes and possibly a cataclysmic event at the West Coast of the US which may be the beginning of a follow up worldwide event which is likely to cause a worldwide cataclysm.

References:

[1]   Albers, C. (2019). Article 992: Huge Planet X object in the sky over Hawaii: causes dizziness and nausea.

[2]   Albers, C. (2019). Article 786: Sky blue Planet X inside the earth's atmosphere.

[3]   Albers, C. (2019). Article 1003: Sinkholes in California close to earthquake swarm region: rift unzipping.

[4]   Albers, C. (2019). Article 997: Earthquake in Seattle on July 12th ground dropping and core reset event.

[5]   Albers, C. (2019). Article 1000: Birch Bay tides connected to Seattle earthquakes and due to Planet X.

Chapter 4

1012. Planet X causes spherical object to emerge from the ground

Figure 1 below shows a spherically shaped protrusion, which appeared on the ground in Beth's property after she sent in the photographs, which indicated that a huge Planet X Stellar Core (SC) was in the sky close to San Francisco (see Article 1005: Planet X in the sky and California earthquakes) [1]. The protrusion looks like a spherical object and it seems to have emerged from the ground. The fact that it is spherical suggests that it is an earth satellite core, which must have been close to the surface and the gravitational interaction between the Planet X SC in the sky and the earth's central core pulled it upwards towards the surface until it emerged.

Figure 4.1. Spherically shaped protrusion emerges from the ground after huge Planet X object observed overhead suggesting that an earth satellite core moved under the surface of the earth upwards until it partially emerged from the ground.

This suggests that Planet X draws earth satellite cores toward the surface and that is thus most likely the reason why core matter, such as trovants, are found on the surface of the earth. Trovants are rocks that grow and reproduce, which identifies them as core matter or matter that creates other matter. Trovants create sand and are thus sand creating cores. The core emerging from the ground in the above image is most likely a soil creating core as well.

Figure 4.2. Trovant babies grow out of a large trovant. Once they reach a certain size the small trovants seem to break off the parent and become independent trovants (see Article 987: Trovants: rocks that grow: core matter, water and human consciousness) [2].

It is not logical that core matter would appear on the surface of a planet, as a planet's core system is what creates the planet's body, unless there is outside interference. That outside interference appears to be Planet X. The gravitational connections it makes with the earth's central core actually seem to draw earth's satellite cores upwards, towards the surface of the earth. This then explains why volcanoes form. These Planet X gravitational connections draw magma creating cores towards the surface, so that magma eventually reaches the surface.

This movement of earth's satellite cores would then be the trigger that would cause increased matter creation at the surface, thus causing the earth to expand, and likely result in the earth's central core ejecting more satellite cores. The movement of earth's satellite cores toward the surface, in response to Planet X, thus seems to be the reason why surface reformation occurs. We are thus seeing surface reformation occurring, on a small scale, with this spherical protrusion and thus earth core, emerging from the ground.

In conclusion, the appearance of a spherical protrusion in the ground after a huge Planet X object, appeared in the sky, suggests that Planet X induces earth satellite cores to move upwards toward the surface of the earth, and that this is the main mechanism leading to surface reformation and the appearance of volcanoes on the surface of the earth.

References:

[1]   Albers, C. (2019). Article 1005: Planet X in the sky and California earthquake.
[2]   Albers, C. (2019). Article 987: Trovants: rocks that grow: core matter, water and human consciousness

Chapter 5

1065. Planet X causes volcano formation by pulling cores to the surface

In Article 1012: Planet X causes spherical object to emerge from the ground [1] I wrote about an object that seemed to be a small soil creating earth satellite core, which emerged from the ground after Planet X planetary central cores, which have become earth moons (see Article 1029: Planet X as new moons orbiting the earth: the irrefutable evidence) [2], were observed in the sky over this location, in northern California. The image was sent in by Beth, who has also now sent me another two photographs after finding two other new rounded mounds indicative of more cores having emerged from the ground in this area.

Figure 5.1. Spherically shaped protrusion emerges from the ground after huge Planet X object observed overhead suggesting that an earth satellite core moved under the surface of the earth upwards, until it partially emerged from the ground. The soil that the object is surrounded with indicates that it is a soil creating core.

Rocks that grow and reproduce, on the surface of the earth, have been known to exist for some time and are called trovants; these would be sand creating cores as they create sand and have a sandstone nucleus.

Figure 5.2. Trovant babies grow out of a large trovant. Once they reach a certain size the small trovants seem to break off the parent and become independent trovants (see Article 987: Trovants: rocks that grow: core matter, water and human consciousness) [3].

Figure 5.3. Beth found this new rounded mound in her garden indicative of a soil creating core having also moved upwards towards the surface and having caused a mound to form as a result.

Figure 5.4. And yet another new rounded mound indicates another core starting to emerge.

This means that Planet X central cores induce the movement of earth cores toward the surface of the earth, which suggests that mountain formation on earth is actually due to Planet X, causing cores to move toward the surface, and if they are deep within a layer of ground, the ground would rise and a mountain would form. And if they are very close to the surface they may actually end up above ground as the trovants have done.

Figure 5.5. Protruding ground formations such as these are thus caused by earth's satellite cores moving upwards due to Planet X rather than the result of erosion.

But even if erosion played a role, the rocky formations, which may become exposed would still have been the result of earth ground creating satellite cores moving upwards toward the surface of the earth in response to a Planet X gravitational connection, with the earth's central core.

We would also expect magma creating cores, which would initially be deep within the body of the earth to move toward the surface, so that eventually the magma finds its way to the surface, and thus a volcano forms, the same would occur in the case of water creating cores, which would give rise to geysers and to oil creating cores which would give rise to tar pits, when Planet X gravitational connections induced them to move toward the surface.

This volcano formation activity from Planet X thus shows which planets in the Solar System are native Solar Planets, since they would have been affected by the presence of Planet X, the core systems of the destroyed Planet X planets and stars, which were destroyed due to energy depletion. Venus has the most volcanoes of all planets in the Solar System, which indicates that it was the planet that was most affected so far, but Mars also has volcanoes on its surface and has thus also been affected.

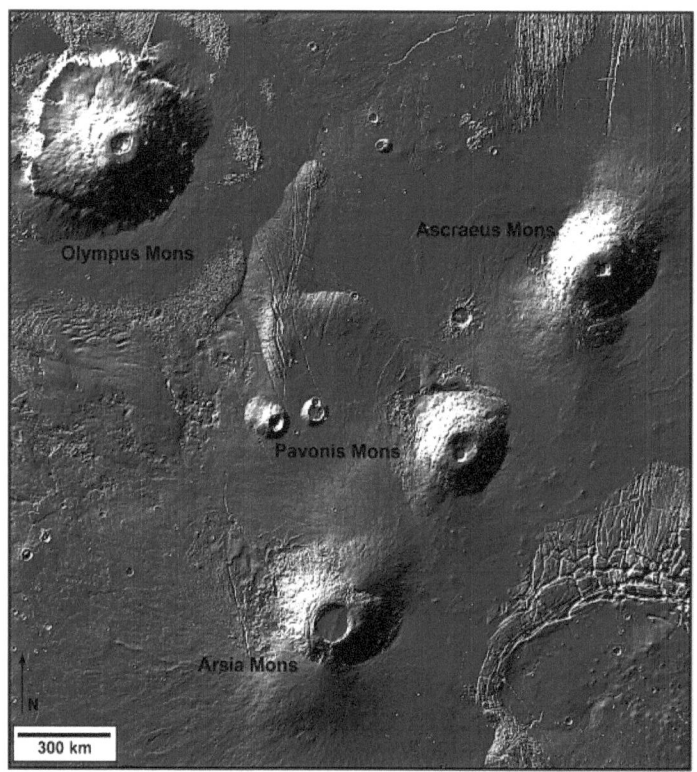

Figure 5.6. Volcanoes on Mars indicate that it is a native Solar System planet, which has been affected by Planet X.

The fact that Venus has become an inferno is however very sobering as clearly more and more volcanoes are forming on the surface of the earth due to ever increasing Planet X systems coming in and this seems to have been part of the process that made life on the surface of Venus impossible and may thus be what will happen to earth as well in the very near future (see Article 1066: Signs of a warming earth due to Planet X: is this what happened to Venus?) [4] The other part of the process was most the increased energy flow from the core which caused the atmosphere and liquids on its surface as this is what is happening on earth due to Planet X.

In conclusion, Planet X through the gravitational connections it makes to the earth's core system causes volcanoes to form by inducing earth cores to move upwards toward the surface of the earth. Through this same process, Planet X also causes the formation of mountains, geysers, tar pits and mud volcanoes, on the surface of the earth, and the existence of volcanoes on other planets, such as Mars and Venus, indicates that Planet X has affected these planets in a similar way that it is affecting earth.

References

[1] Albers, C. (2019). Article 1012: Planet X causes spherical object to emerge from the ground.

[2] Albers, C. (2019). Article 1029: Planet X as new moons orbiting the earth: the irrefutable evidence.

[3] Albers, C. (2019). Article 987: Trovants: rocks that grow: core matter, water and human consciousness.

[4] Albers, C. (2019). Article 1066: Signs of a warming earth due to Planet X: is this what happened to Venus?

Chapter 6

1082. Planet X effects on Mars: volcanoes and craters

Mars, the planet in the Solar System which is a bit further out from the Sun than earth seems to have been deeply affected by Planet X, as it has both volcanoes and huge craters, which are really sinkholes. Both are known to form as a result of gravitational connections between Planet X central cores and a planet's central core. Volcanoes form because Planet X pulls cores toward the surface of the planet. Thus, when gravitational connections with magma creating cores occur these are pulled toward the surface until a volcano forms at the connection location (see Article 1065: Planet X causes volcano formation by pulling cores to the surface) [1]. Craters form in the connection region between two central cores as well.

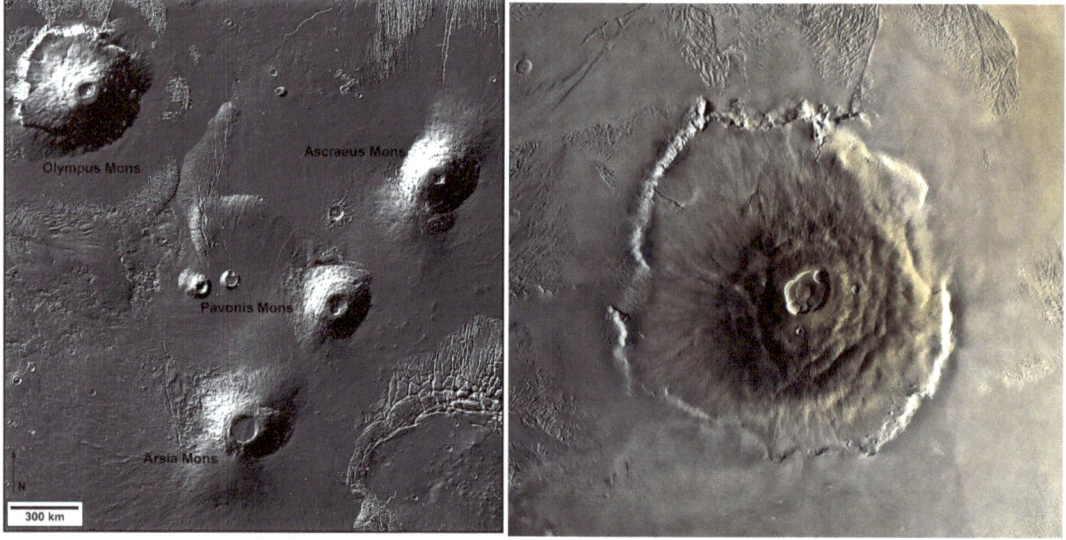

Figure 6.1. Olympus Mons is the largest volcano on Mars and in the Solar System as it is 22 km (12.4 miles) high. The other three are also huge, over 10 km in height. Olympus Mons is 640 km (397 miles) wide.

The size of these huge volcanoes suggests that Mars was affected by a very large Planet X central core, which had an extremely destructive effect on the planet. Mars also has huge craters; the Lyot Crater, which is 236 km (147 miles) in diameter and the Lowell crater, which is 203 km (126 miles) in diameter. Both are circular and symmetrical with extra inner rings and thus the result of a gravitational connection producing a gravitational interference pattern, which results in the creation of sinkholes on the surface. The object responsible would have to be larger than 236 km in diameter in order to be able to make a gravitational connection, which resulted in a sinkhole of the size of the Lyot Crater.

Figure 6.2. Left: Lyot Crater (236 km in diameter): it has two raised rings, one forming outer perimeter of the sinkhole, 2 lower elevation regions and a central peak. **Right:** Lowell Crater (203 km in diameter): it has two raised rings, one forming the outer perimeter and two lower elevation regions, one being the central region.

Thus, the gravitational interference pattern produced at the Lyot Crater had a central maximum, which produced the central peak, but the interference pattern, which created the Lowell Crater, had a central minimum, which created a central hollow. Other much smaller craters formed both on top and around the two gigantic craters. The smaller craters have an outer raised ring at the perimeter and a central hollow, with some having raised central peaks as well, but all are circular and symmetric and thus not created through direct impact but due to gravitational connections, without the Planet X central cores ever making contact with the surface of Mars.

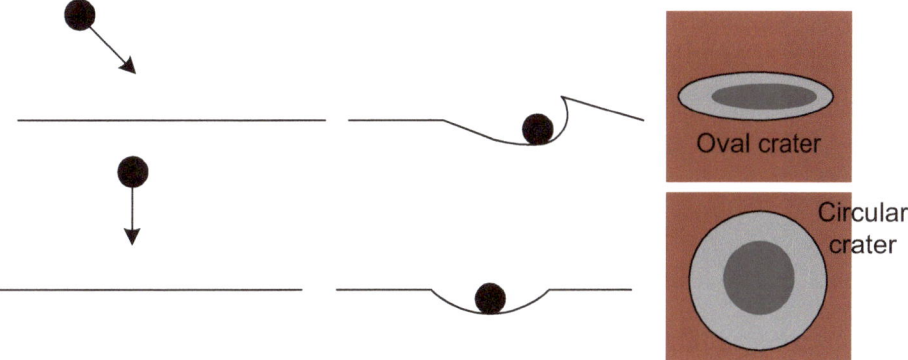

Figure 6.3. An impact crater cannot be expected to be circular in shape unless the impacting object comes in at an angle of 90°. But all objects in the universe follow curved trajectories around a massive object and will thus never come in at right angles, making a circular impact crater impossible. A real impact crater will always be oval and deeper on one side. Thus, circular craters are not impact craters, they are sinkholes produced as a result of gravitational interference (see Article 675: Sinkholes all over the earth: impact craters are sinkholes) [2].

Planet X cores when making gravitational connections become a part of a planet's core system and are thus able to remain stationary with respect to its surface and to also slowly move vertically downwards toward the surface, until they get to a minimum distance where they are repelled by the surface. That is why they able to cause the formation of circular craters without ever impacting the surface of a planet. As to craters being due to an electrical interaction, Planet X cores absorb energy from the earth's core in the form of static electricity and are thus capable of creating lightning discharges and the larger the objects the larger these discharges are likely to be, but sinkholes often form without any lightning being visible, so it is not a necessary condition.

Figure 6.4. The formation of a sinkhole: The sinkhole forms under the surface, there is no direct force from the top, the ground under the surface moves downwards first, then the surface of the road falls inwards due to having lost the ground's support. This shows that the interaction leading to the formation is between the Planet X central core and the earth's central core, resulting in the earth moving the position of the earth's surface in response to the presence of the Planet X central core. A dome can be seen in the center of the hole, showing that this sinkhole formed as a result of a gravitational interference pattern with a central maximum (see Article 674: Sinkhole in Knoxville Tennessee due to Planet X and gravitational interference) [3].

In conclusion, Mars seems to have been deeply affected by Planet X as it is covered in huge volcanoes and craters, which are really sinkholes, some of which are also huge in size, having a multiple ring structure, due to the gravitational interference, which must have occurred between a large Planet X central core and Mars' central core.

References:

[1] Albers, C. (2019). Article 1065: Planet X causes volcano formation by pulling cores to the surface.
[2] Albers, C. (2019). Article 675: Sinkholes all over the earth: impact craters are sinkholes.
[3] Albers, C. (2019). Article 674: Sinkhole in Knoxville Tennessee due to Planet X and gravitational interference.

Chapter 7

1134. Intelligent life all over the Universe

Planet X, the core systems and debris fields of previously destroyed star systems, which were destroyed as a result of energy depletion and thus coming into the Solar System to absorb energy, show that there were once many star systems in the Milky Way Galaxy that contained intelligent flesh and blood beings like us and that they were technologically advanced. All these systems were however destroyed because of Lucifer, the Destroyer, who filled the minds of the inhabitants with destructive thoughts, which then led to Planet X coming in and destroying one system after another (see Article 1133: The beginning of all the destruction in the universe) [1].

Figure 7.1. Cities in the sky: The same city in the sky seen in China in 2017, photographed from different angles. These are buildings on Planet X surface debris pieces, coming into the earth's atmosphere (see Article 584: Planet X the reason behind GMO crops) [2].

Figure 7.2. Floating city in the sky: There also seems to be greenery around the buildings, indicating that whatever seeds, were on this piece of a broken planet, were still viable and have germinated most likely after entering the earth's atmosphere as plants need carbon dioxide and an atmosphere to grow. This shows why insect eggs hatch and produce insect plagues (see Article 798: Insect plagues: Planet X surface debris coming in slowly) [3] as well as the viruses and parasites that were created to be biological weapons by Lucifer, first of all, and then modified and reused on the Planet X planets [1].

Figure 7.3. Rocks (jagged edges like mountain tops but disconnected from the ground) or part of one of the Planet X planets' debris field suspended in the earth's atmosphere: The planets were destroyed due to energy depletion, so the rocks have the same gravitational energy as material, which on earth, is only found in the atmosphere, due to their low density. The Planet X rocks absorb energy slowly and thus sink toward the surface of the earth, slowly. The larger the rocks the slower they absorb energy and thus the slower they will sink, which therefore shows why an asteroid, since these are all Planet X debris, will never impact the earth and cause a huge explosion.

Figure 7.4. A huge rock appears amongst the clouds in Peru: This piece of rock is long and flat like the ones with buildings on them, which suggests that it is a surface piece, and that the surface of the planets broke into long flat pieces of rock.

Figure 7.5. Buildings within clouds are seen floating in the sky, over the ocean. The buildings are just like the buildings found on earth and thus suggest that the intelligent beings on this planet were very similar to us and also had similar technological capabilities. The buildings seem to be at a lower altitude than other cities observed in the sky and them being over the ocean suggests that these are purposely steered to the ocean, so that they end up sinking below the waves and disappearing.

Figure 7.6. A view from an airplane (from a MrMBB333 video) reveals another piece of Planet X surface debris with buildings and other artificial structures on it. A straight line can be seen close to the left edge, indicating the presence of a building or wall, and a dome appears to emerge from the center of the circular structures. There are many small light colored structures within the image that do not appear to be cloud formations, there is also a cross, which may have been antenna, and two bent cylindrical structures aligned with each other, as if they were meant to be one structure with a gap in it (see Article 975: Rocks with buildings and antenna suspended in the air: Planet X debris) [4].

Figure 7.7. Island appears in the middle of a frozen lake, where no island had ever been, in Finland. No volcanic or seismic activity could explain it either. The island seems to be bordered by an artificial structure, which looks like columns, all of the same height, and neatly aligned, which cannot in any way be natural. The columns also have colored geometric patterns on them (see Article 642: New Island with artificial structures: Planet X destroyed planets' debris) [5]. And no, this is no mirage; a mirage is the product of the human mind trying to make sense of something, which it cannot clearly see. If this had been a mirage, it would not be clearly defined, we would be wondering what it was and applying our imagination, but no imagination is necessary here; artificial columns and square patterns in different colors can be clearly discerned. This must be a piece of the Planet X debris, with artificial structures on it, which has slowly sunk in the earth's atmosphere, and was, most likely, steered to this lake, by the 'powers that be', so that it would quietly sink into it. The white patches close to it suggest that it had started breaking through the ice. These structures may well have been contaminated with insects or viruses used in the war that was surely waged on these planets by Lucifer.

Figure 7.8. Signs of a destroyed civilization on Mars: a human face and pyramids, which appear to be in ruins. The human face tells us what the inhabitants looked like and that they were thus very much like us (see Article 1087: Nuclear war leads to Planet X cataclysm on Mars and on Earth) [6].

Hence, Planet X research leads to the conclusion that God created a universe where planets naturally form with habitable environments designed for animals and flesh and blood beings, like we have on Earth, and that it is therefore God's will to fill the universe with intelligent beings like us. It was therefore from the beginning of the creation of the physical universe, God's intention, that humans spread throughout the universe and inhabit the planets He had created. Lucifer interrupted God's plan but God has not changed His mind and once Lucifer is contained, His plan to fill the universe with intelligent life will proceed. This explains why most human beings look at the stars with longing, He has built within us, the desire to go out there and explore the universe and make it our home.

God has also created humans to procreate or multiply and even the way that procreation is done has God's signature upon it, because the connection between male and female is similar to gravitational connections between cores and which we see in the case of a tornado core making a gravitational connection with the earth.

Figure 7.9. Tornado creating gravitational connection: Planet X core or Stellar Core (SC) makes what is usually called a male plug and the earth material in the outer ring region a female plug.

In conclusion, the physical universe God has created is not meant to be empty, we are supposed to go out and explore it and inhabit the planets throughout this galaxy and beyond.

References:

[1] Albers, C. (2019). Article 1133: The beginning of all the destruction in the universe.
[2] Albers, C. (2019). Article 584: Planet X the reason behind GMO crops.
[3] Albers, C. (2019). Article 798: Insect plagues: Planet X surface debris coming in slowly.
[4] Albers, C. (2019). Article 975: Rocks with buildings and antenna suspended in the air: Planet X debris.
[5] Albers, C. (2019). Article 642: New Island with artificial structures: Planet X destroyed planets' debris.
[6] Albers, C. (2019). Article 1087: Nuclear war leads to Planet X cataclysm on Mars and on Earth.

Chapter 8

1140. Spherical objects and soil emerging from ground in California due to Planet X

In Article 1012: Planet X causes spherical object to emerge from the ground [1], I detailed the emergence from the ground of a spherical object, which could only have been an earth core. The event was noticed by Beth, who lives in northern California and first sent me photographs of the sky over her location, which indicated the presence of huge Planet X cores or Stellar Cores (SCs), which were then creating sparse water vapor clouds on their surfaces. These photographs are from July of 2019 and appear in Article 1005: Planet X in the sky and California earthquakes [2]. Soon after that she noticed that mounds were forming on her property and it looked like spherical objects were emerging from the ground.

Figure 8.1. Spherically shaped protrusion emerges from the ground after huge Planet X object observed overhead suggesting that an earth satellite core moved under the surface of the earth upwards, until it partially emerged from the ground. The soil that the object is surrounded with indicates that it is a soil creating core.

She then found more of these mounds and has now in September of 2019 found even more, so it seems that earth cores are increasingly emerging from the ground at this location. Planet X observations show that all celestial objects have core systems in them and that the cores are sources of gravity and matter and thus the emergence of earth cores from the ground suggests that this will happen with all earth cores with which energy depleted Planet X cores connect with, in order to absorb energy, and that this is the process, which then causes volcanoes to form (see Article 1065: Planet X causes volcano formation by pulling cores to the surface) [3].

Figure 8.2. Beth found this rounded mound in her garden (August of 2019) indicative of a soil creating core having also moved upwards towards the surface and having caused a mound to form as a result.

Earth cores have been on the surface of the earth for quite some time such as the trovants, shown below, which grow and reproduce and would also have initially emerged from inside the body of the earth due to Planet X.

Figure 8.3. Trovant babies grow out of a large trovant. Once they reach a certain size the small trovants seem to break off the parent and become independent trovants (see Article 987: Trovants: rocks that grow: core matter, water and human consciousness) [4].

Figure 8.4. Photograph sent by Beth on September 19th 2019 of another earth soil creating core emerging from the ground, newly formed soil seems to be emerging as well suggesting that the core is creating the soil as Planet X cores, in the sky, connect with it and absorb energy from it.

Figure 8.5. Another photograph showing the soil which is appearing: The soil seems to be below the vegetation and thus emerging from below, and must thus, be newly created by the emerging earth soil creating cores.

The earth cores emerging from the ground seem to be alive, as they are creating soil in response to gravitational connections with Planet X cores, but there are many earth cores, found all over the earth, which are no longer alive as they do not seem able to create new matter anymore, i.e. they are no longer able to absorb energy in order to then create matter, so they have turned into dead matter or to created matter rather than creating matter.

Figure 8.6. Spherical stones are found all over the world: **Top left**: Bosnia. This stone has a 10 foot diameter and appears to have a high iron content. **Top right:** New Zealand. **Bottom left:** Mexico. **Bottom right:** Costa Rica. These have to be earth cores, which have emerged from the ground due to Planet X, but they no longer create matter, which suggests that they are dead.

Even the trovants do not seem to just grow matter, without being exposed to rain, which suggests that they absorb energy through the rain water. Planet X cores in water creating phase cause it to rain so that energy flows from the earth's core system through the rain water to them. It is thus possible that these spherical cores would also start creating some matter when exposed to rain.

Figure 8.7. Beth also noticed that holes have appeared in the ground, where the newly created soil has emerged and where the presence of a new mound would indicate that an earth core is moving toward the surface.

The holes would have to thus be the result of the gravitational connections, which Planet X cores are making with this earth core. These would then be small sinkholes and thus suggests that sinkholes can form at any level of gravitational connections, i.e. not just when Planet X cores connect to the earth's central core. This would then suggest that the mouth of a volcano is actually a sinkhole, produced by the gravitational connection between a Planet X core and an earth core. This would then agree with a conclusion I reached in Article 1006: Volcanoes and sinkholes: created through the same process [5].

Sinkholes seem to form in the center of the connection and are due to the attractive force, which an earth core exerts on a Planet X core, which then causes it to drop material it creates, in the connection region. I have named this cloud spout material before, but it only applies when the cores are in water creating phase, when in other phases, such as oil or magma creating phase, calling it energy transfer conduit material would make it more applicable to other matter creation phases. Sinkholes would, however, only form when the Planet X core in the sky is smaller than the earth core below the ground.

Figure 8.8. The gravitational connection has at least two regions, a central region, where the smaller core's created matter moves toward the larger core, and an outer ring where the larger core's matter moves toward the smaller. Here the SC is smaller, so cloud spout material moves down in the central region and earth soil moves upwards in the outer ring region.

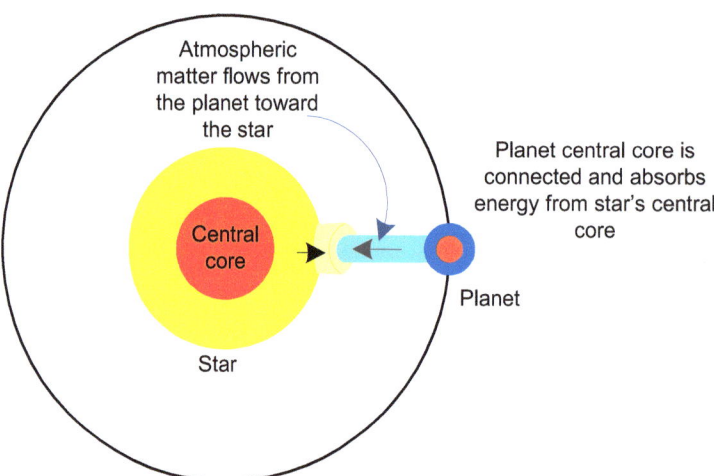

Figure 8.9. Gravitational connection between a planet and a star: The planet drops created atmospheric matter, in the central region, which acts as an energy conduit, for the energy from the star's central core to move through. The denser the matter in the connection region, the more energy flows across to the planet, which it can then use to draw more energy. If a planet's ability to draw energy drops, it is less able to draw energy from the star. This is the process that leads to cores dying, i.e. becoming so depleted that they are no longer able to create matter, in order to draw energy from other cores.

The reason why earth cores would move closer to the surface of the earth and thus further from the earth's central core is because of a drop in gravitational energy. Because they are transferring energy to Planet X cores, as well as creating matter in response to the gravitational connections, their own energy drops, and this would cause them to move away from central core. This is a normal process in the universe and is what turns a globule of plasma ejected by a galactic nucleus into a globular cluster of individual stars. This is also the process, which have occurred in the Planet X planets as previously destroyed systems came in to absorb energy and which ended with the planets breaking up.

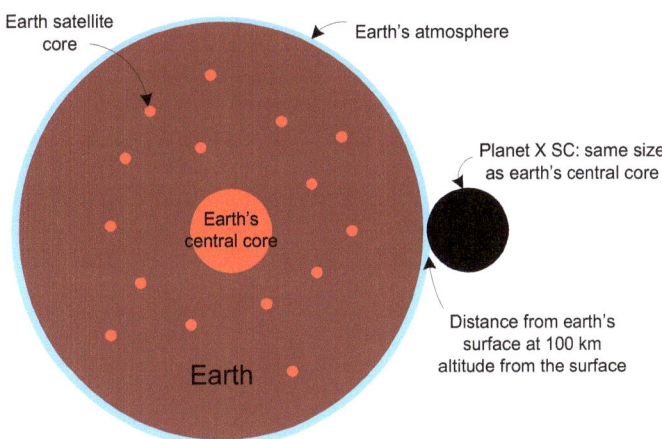

Figure 8.10. A core system is made up of a central core and many much smaller satellite cores, which were ejected by the central core after it emerged from the body of the star, it was ejected by. Finding itself in a much lower gravitational potential region, the newly ejected core goes into massive matter creation mode and ejects its own satellite cores, which then create matter and a planetary body and atmosphere forms around the central core.

But as a planet's gravitational energy is quickly absorbed to the point that its central core becomes depleted, the whole planet breaks into pieces, its satellite cores would then end up in space far from the central core and amongst all the debris that used to be the body of the planet. They would then be like any created matter but still able to absorb energy once inside the atmosphere of a star or planet with a living core system.

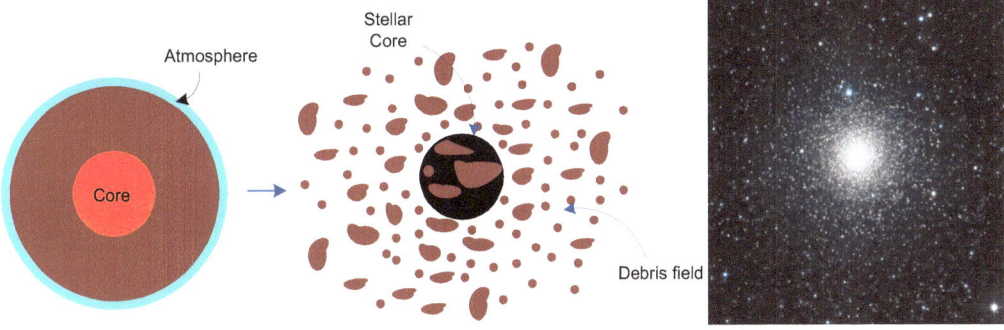

Figure 8.11. **Left:** Planet X planets became a core system with a central core and much smaller satellite cores (not shown) surrounded by debris pieces due to being affected by previously destroyed star systems coming in to absorb the energy of the planet as is now happening with earth. When the central core's gravitational energy becomes depleted the whole planet expands outwards turning into a debris field of broken pieces of planet. The suddenness of the event is apparent in the surface pieces coming into the earth's atmosphere that still have buildings on them (see Article 584: Planet X the reason behind GMO crops) [6]. **Right:** NGC6388 Globular Cluster, in the Milky Way Galaxy: It is made of separate stars and the star density increases toward the center, where the gravitational energy is higher. The gravitational energy continuously drops due to continuous matter creation causing the created matter to fission and move further away from the globular cluster core.

Thus the emergence of cores from the ground shows that they are losing gravitational energy due to Planet X drawing energy from them. And the fact that this seems to be happening more and more suggests that the process is accelerating. In addition, when these same Planet X cores now drawing energy from these earth creating cores, reach the magma creation phase, it will not be soil and soil cores that will emerge from the ground, but magma as new volcanoes form. This seems to be what happened on Venus on a very large scale (see Article 1117: Planet X destroyed Venus and is now destroying Earth) [7].

Both Venus and Mars seem to have been deeply affected by Planet X, to the point that they are no longer able to support life, but they did not break up because satellite cores, at a certain level in their core system, became completely energy depleted before the central core could lose enough gravitational energy for the whole planet to break apart. The dying of cores at a certain level then stops Planet X cores from proceeding to the next level to absorb energy (see Article 1135: Planet X research agrees with Genesis 1: earth recreated) [8]. However, all satellite cores above that dead level would most likely also become completely energy depleted, before Planet X cores would move away for the planet, leaving the top part of the planet in a dead state, unable to create soil or water, close to the surface. This explains why the Sun became permanently dark around 1998 (see Article 536: Was the Sun already not shining by 1998?) [9], when its light creating cores died but the Planet X cores did not start coming to earth in larger and larger numbers until more recently (see Article 1084: Planet X over Texas emits light: dark Sun signaled the end for earth and Article 1056: Darkness instead of daylight again in Siberia due to Planet X) [10, 11]. The Sun must have still had cores with energy in them, in the levels above the light creating cores, which the Planet X cores continued to deplete until they reached the light creating level and only left the Sun to seek another source of energy, then, and thus coming to the one planet that still has a completely intact core system, the Earth. The reason why the Solar System are not exploding like the Planet X celestial objects is most likely because the satellite cores in the Solar System objects, closer to the surface, had less energy in them from the beginning.

In conclusion, spherical objects and new soil are emerging from the ground and also tiny sinkholes are appearing in California, due to Planet X making gravitational connections and absorbing energy from the Earth's core system, in a process, which led to the Planet X planets breaking up, and that will also destroy the earth, even if not to the extent that it breaks up as well.

References:

[1] Albers, C. (2019). Article 1012: Planet X causes spherical object to emerge from the ground.
[2] Albers, C. (2019). Article 1005: Planet X in the sky and California earthquakes.
[3] Albers, C. (2019). Article 1065: Planet X causes volcano formation by pulling cores to the surface.
[4] Albers, C. (2019). Article 987: Trovants: rocks that grow: core matter, water and human consciousness.
[5] Albers, C. (2019). Article 1006: Volcanoes and sinkholes: created through the same process.
[6] Albers, C. (2019). Article 584: Planet X the reason behind GMO crops.
[7] Albers, C. (2019). Article 1117: Planet X destroyed Venus and is now destroying Earth.
[8] Albers, C. (2019). Article 1135: Planet X research agrees with Genesis 1: earth recreated.
[9] Albers, C. (2019). Article 536: Was the Sun already not shining by 1998?
[10] Albers, C. (2019). Article 1084: Planet X over Texas emits light: dark Sun signaled the end for earth.
[11] Albers, C. (2019). Article 1056: Darkness instead of daylight again in Siberia due to Planet X.

Chapter 9

1144. The Universe is alive: God is reproducing Himself

Beth started sending me photographs of huge Planet X cores in the sky, over her location in California, in July of 2019. Soon after that, she noticed that new mounds were forming on her property and that spherical objects were emerging, which also created soil. These turned out to be earth cores or a part of the earth's core system (see Article 1140: Spherical objects and soil emerging from ground in California due to Planet X) [1]. Figure 1 below shows a photograph sent to me by Beth of a small core, which has completely emerged from the ground on her property. This core seems to be smaller than the others and was stepped on, so the inside can be seen.

Figure 9.1. Left: Tiny earth core, which seems to have completely emerged from the ground: The inside appears to be made of dark soil and it has a grey outer layer. The amount of grey soil around the object would suggest that this is the kind of material it creates as an envelope or skin. The core is obviously very small and fragile. **Right:** Planet X core or Stellar Core (SC) in the earth's atmosphere, obviously much larger and most likely made of much denser matter as it appears to be spherical. This core would have been a satellite core in one of the Planet X planets.

The color of the soil around the small core which matches its envelope shows that the soil created by this core is part of its envelope. Small soil creating cores would be at the final edge of what core matter creates in the universe. In other words, these would be the most specialized and lowest energy core matter generated by the universe.

Figure 9.2. Earth cores which are made of much denser matter and also spherical in shape like the SC in the earth's atmosphere.

Figure 9.3. Trovant babies grow out of a large trovant. Once they reach a certain size the small trovants seem to break off the parent and become independent trovants (see Article 987: Trovants: rocks that grow: core matter, water and human consciousness) [2]. These are earth cores which have also emerged from the ground. They create envelope matter and also reproduce indicating that cores are alive. The babies have the same internal structure and outer envelope as the parent.

The earth's central core would have ejected the small soil creating core, in figure 1, as well as many others, when the planet first formed. Each core would grow and be ejected from a specific location on the central core which would have a unique composition, as the central core would be made up of a huge quantity of different materials, each satellite core would be made of only some of these materials and the smaller the satellite core the more specialized it would become so that it would only be able to create babies with what it was composed of. The trovants are very specialized as they only create one type of baby and so would the small soil creating cores like the one in figure 1. Once ejected, the satellite cores would move out from the core to the position, which would be determined by their energy potential or gravitational potential, the smallest earth cores would have the lowest energy potential, and thus, would end up furthest from the central core. The number and material composition of these soil creating cores would then determine what type of surface the planet would end up with.

Thus, the universe is designed to end the creation process, which starts with matter ejected by a galaxy, with the formation of unique planets.

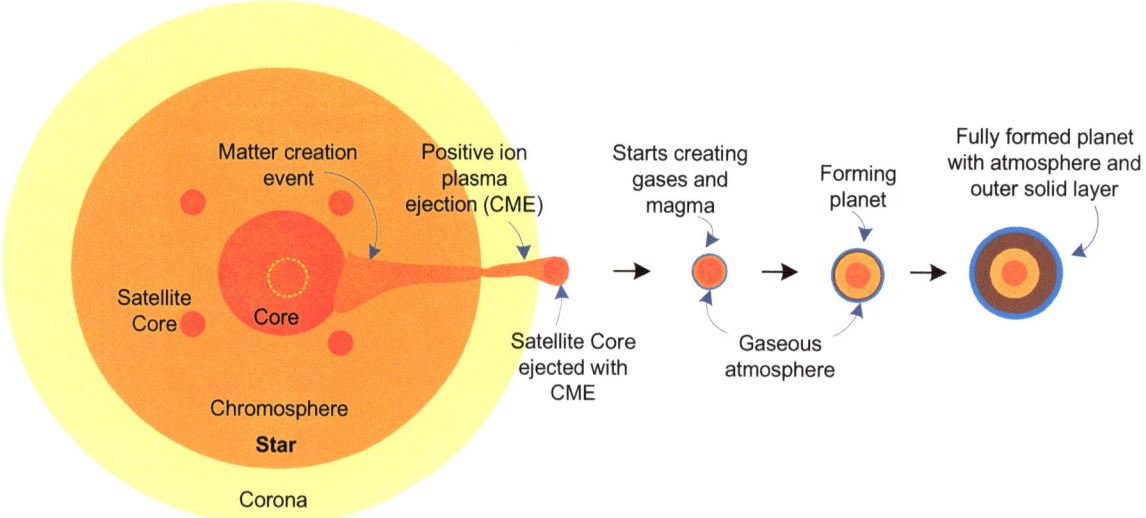

Figure 9.4. A star creates or grows and ejects satellite cores with different compositions, some of which end up outside the body of the star. Once outside the body of the star, the new star satellite core, finding itself in a lower energy environment, creates an envelope with which to absorb matter from the environment, or the star, it has left, by growing and ejecting small parts of its own core, which become its satellite cores, these then create their own energy absorbing envelopes by creating matter, which turns into the body of the planet and the central core's envelope. The smallest cores, will have the least energy and end up closest to the surface, these will tend to be made of one specific material and create one type of soil as an envelope.

Planetary central cores, just like star central cores, grow or create satellite cores on different points on their surface which have a different composition, so that the babies have different compositions. But the smallest cores, i.e. soil creating cores which end up closest to the surface of the planet, seem to be more specialized and some even fragile like the soil creating core in figure 1, which suggest that these soil cores are the end products of the matter creation, in the universe, which started with a galactic core reproducing, i.e. creating stars. Thus, all celestial objects grow and reproduce and are thus alive. The universe thus seems to be saying that it is designed to reproduce because God is reproducing Himself in flesh and blood human beings, who also reproduce. Each human being has a number of God's characteristics, just like the satellite cores have some of the compounds, which the central core contains. God's plan, before all the Planet X planets were destroyed, which were inhabited by beings like us, and now, is that the planets which the universe is designed to create will eventually be inhabited by beings like us across the different galaxies (see Article 1134: Intelligent life all over the Universe) [3].

In conclusion, matter creation in the universe seems designed to create unique planets. In addition, all celestial objects in the universe, from galactic nuclei to planetary cores grow and reproduce and are thus alive. Core matter seems to also reproduce in the same way that God reproduces Himself in us, as we each have a few of his characteristics just like satellite cores have some of the materials contained in the central core but not all of them.

References:

[1] Albers, C. (2019). Article 1140: Spherical objects and soil emerging from ground in California due to Planet X.

[2] Albers, C. (2019). Article 987: Trovants: rocks that grow: core matter, water and human consciousness.

[3] Albers, C. (2019). Article 1134: Intelligent life all over the Universe.

Chapter 10

1147. Planet X effects accelerating: water emerges in California

I have now written several articles detailing the emergence of earth cores from the ground, based on photographs sent to me by Beth since July of 2019. So far several earth soil creating cores have been observed emerging from the ground on Beth's property, in California. These earth cores create soil, which is thus observed emerging from the ground (see Article 1140: Spherical objects and soil emerging from ground in California due to Planet X) [1]. But now Beth has also found evidence of water emerging from the ground.

Figure 10.1. Left: A clearly spherical object emerges from the ground, the new soil, which also has emerged identifying it as a soil creating core. **Right:** A new mound covered in new emerging soil indicates that another core is emerging at this location. Some of the soil is likely to have been created and dropped by the Planet X core making a gravitational connection with the earth core.

Figure 10.2. After cutting away at a vine, Beth found underneath it newly emerged soil and also wet soil. Since it has not rained, the water and new soil can only mean that both a soil creating core and a water creating core are emerging at the location.

Figure 10.3. In this image we can see the tiny sinkholes which the gravitational connections between Planet X cores, in the sky, and earth cores, give rise to. The newly emerging soil is also very compact and darker in places, a sign that it is wet. This would suggest that a water creating core has been reached by the Planet X cores in the sky and is now also ejecting water and approaching the surface.

The gravitational connections would automatically cause earth cores to move closer to the surface because their gravitational potentials would drop as they create matter and transfer energy to the Planet X cores. As Planet X cores gain energy they are expected to make connections with earth cores deeper and deeper, within the body of the earth, which will then also start ejecting matter and moving toward the surface.

Figure 10.4. The huge amount of new soil in this photograph suggests an acceleration in the process at this location, most likely because more and more soil creating cores are being reached and activated by Planet X cores, coming in to this location to make gravitational connections, and to draw energy from the earth's core system. At least some of the soil seems to have been dropped rather than have emerged from the ground which suggests that it was created by the Planet X core making the connection to an earth core under the surface. Some of the dropped soil is in the form of hard clumps.

The body of the earth is the central core's envelope, so as Planet X extracts energy from its satellite cores, they are, in fact extracting energy from the earth's central core and changing the earth as a result.

This scenario at this location in California is most likely being repeated in many locations around the world. It is a process, which is likely to eventually cause earthquakes as deeper cores are reached and start their ejection and movement toward the surface, and eventually lead to the formation of volcanoes (see Article 1065: Planet X causes volcano formation by pulling cores to the surface) [2].

Figure 10.5. August 28th 2019 eruption of the Stromboli Volcano from a boat showing the ash cloud moving over the ocean close to the island (**Source:** Volcano Discovery). The ash cloud will contain some water vapor as well as soil particles and thus this core, which must be much larger than the soil creating cores and much more energetic creates both water and soil. After the ash cloud is ejected, there is often incandescent gaseous material and after that liquid magma, or dense extremely hot rocky soil in a liquid state. Thus the core ejects the same materials as what is in the initial gaseous cloud but these get hotter and denser indicating an increase in the energy output of the core, as the ejection progresses. Material ejected by a magma creating core is its envelope matter, just like soil is envelope matter for a soil creating core.

Just like soil creating cores come in different sizes, we would expect water and magma creating cores to also come in different sizes but would also expect the average size of magma creating cores to be much larger than that of water creating cores and soil creating cores. Soil ejected by soil creating cores is obviously cool but water coming from geysers is hot and magma from volcanoes is even hotter, which suggests that the hotter the matter ejected by cores the deeper their original positions within the earth.

Figure 10.6. A geyser: A water core, which has clearly been induced to get very close to the surface is seen ejecting hot water and water vapor clouds. If this core was much deeper in the earth, the water would have cooled down by the time it reached the surface. Thus, the water core under Beth's property must still be very deep underground or we would have had hot bubbling water emerging from the ground.

In conclusion, the emergence of soil cores, increasing amounts of soil and the signs suggesting that a water creating earth core is emerging, at a location in California, suggests that Planet X effects on the earth are accelerating. These effects will eventually lead to large portions of the earth's surface being destroyed and the formation of new volcanoes, which will be increasingly cataclysmic.

References:

[1]   Albers, C. (2019). Article 1140: Spherical objects and soil emerging from ground in California due to Planet X.

[2]   Albers, C. (2019). Article 1065: Planet X causes volcano formation by pulling cores to the surface.

Chapter 11

1150. Earthquakes: Planet X effect on earth's cores: not tectonic plate movement

Figure 1 below shows several earthquakes that have occurred in the last few hours in North America on September 24th 2019. Earthquakes are usually explained as being due to movement of the earth's crust along tectonic plates or along faults, which are large cracks in the earth's crust. However, as I have shown in Article 1142: Cataclysmic surface reformation starting due to Planet X: earthquakes and Article 1143: Earthquake in Albania: earth's core system affected by Planet X [1, 2] earthquakes are due to the movement of earth cores, toward the surface of the earth, due to Planet X making gravitational connections and extracting energy from the earth's core system. The cracks or faults and tectonic fractures are caused by this movement of the earth's core system due to Planet X.

Figure 11.1. Earthquakes in North America in the last few hours on September 24th 2019.

It is typical of the aliens who have filled humanity's knowledge base with lies to attribute cause to what is really a consequence (see Article 1040: Aliens have filled humanity's knowledge base with lies and Article 1138: Planet X cover up: Aliens destroying humanity and Earth with lies) [3, 4]. The cause is Planet X, the shaking and cracking of the earth's crust is the result. In fact, the earth can be seen to crack, i.e. fissures open up, when there is an earthquake.

Figure 11.2. A huge crack opens up in Pakistan due to an earthquake that was initially reported as over a 6 magnitude but has since been downgraded to a 5.6 on September 24th 2019.

Figure 11.3. Huge cracks opened in the ground because earth cores moved closer to the surface. This causes the ground to shake and expand, on the surface, which then causes it to crack open.

Planet X gravitational connections cause cores which form a part of the earth's core system to lose gravitational energy and thus move away from the central core, or closer to the surface of the planet, just like a satellite that loses energy, i.e. speed, would move to a higher orbit, gravitational energy increases with decreased distance to the center of a planet.

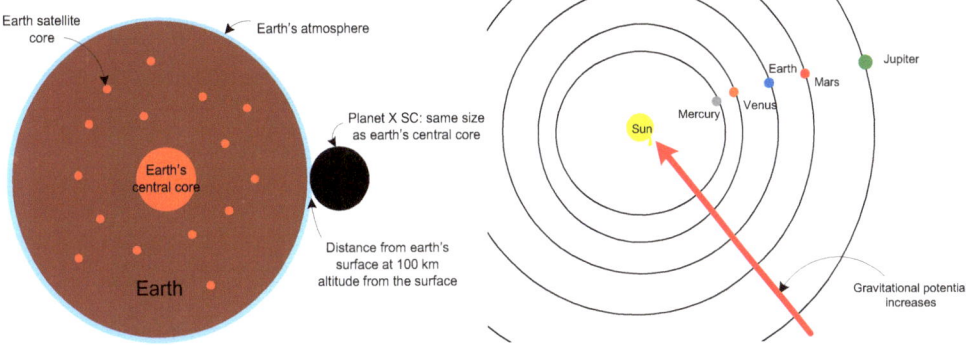

Figure 11.4. Satellite cores' position from the central core depends on their gravitational potential, which drops as they become energy depleted causing them to move away from the central core, just like planets would move to higher orbits if their gravitational potentials drop.

Figure 11.5. One of the cores, which just popped up on Beth's property soon after huge Planet X cores were observed overhead (see Article 1140: Spherical objects and soil emerging from ground in California due to Planet X) [5].

Figure 11.6. Left: Tiny earth core, which seems to have completely emerged from the ground: The inside appears to be made of dark soil and it has a grey outer layer. The amount of grey soil around the object would suggest that this is the kind of material it creates as an envelope or skin. The core is obviously very small and fragile. **Right:** Planet X core or Stellar Core (SC) in the earth's atmosphere, obviously much larger and most likely made of much denser matter as it appears to be spherical. This core would have been a satellite core in one of the Planet X planets (see Article 1147: Planet X effects accelerating: water emerges in California) [7].

Figure 11.7. Trovant babies grow out of a large trovant. Once they reach a certain size the small trovants seem to break off the parent and become independent trovants (see Article 987: Trovants: rocks that grow: core matter, water and human consciousness) [6]. These are earth cores which have also emerged from the ground. They create envelope matter and also reproduce indicating that cores are alive. The babies have the same internal structure and outer envelope as the parent.

There are many types of cores of many different sizes and creating many types of different materials inside the body of the earth. Water creating cores seem to generally be larger and more energetic than soil creating cores and magma creating cores seem to be even larger and more energetic and were thus most likely originally deeper within the planet than soil or water creating cores (see Article 1147: Planet X effects accelerating: water emerges in California) [7]. The movement of magma creating cores toward the surface of the earth eventually leads to the formation of volcanoes.

Figure 11.8. The volcanic cloud starts out looking white, grey cloud material follows, and finally, orange or peach colored cloud material is seen emerging from the Stromboli Volcano during the huge eruption of July 3rd 2019. This suggests that water vapor emerges first and then rocky compounds in increasingly denser form follow (see Article 974: Huge eruption from the Stromboli Volcano: Planet X cataclysm imminent) [8].

In conclusion, earthquakes are due to Planet X's energy depletion effect on the earth's core system, which causes earth satellite cores to move toward the surface and in the process cause the formation of volcanoes. This movement causes cracks to appear, in other words, faults which are simply large cracks and tectonic plate boundaries which are even larger cracks are the result of this movement and therefore a consequence and not the cause behind earthquakes. This process is what also led to the Planet X planets breaking into pieces.

References:

[1] Albers, C. (2019). Article 1142: Cataclysmic surface reformation starting due to Planet X: earthquakes.

[2] Albers, C. (2019). Article 1143: Earthquake in Albania: earth's core system affected by Planet X.

[3] Albers, C. (2019). Article 1040: Aliens have filled humanity's knowledge base with lies.

[4] Albers, C. (2019). Article 1138: Planet X cover up: Aliens destroying humanity and Earth with lies.

[5] Albers, C. (2019). Article 1140: Spherical objects and soil emerging from ground in California due to Planet X.

[6] Albers, C. (2019). Article 987: Trovants: rocks that grow: core matter, water and human consciousness.

[7] Albers, C. (2019). Article 1147: Planet X effects accelerating: water emerges in California.

[8] Albers, C. (2019). Article 974: Huge eruption from the Stromboli Volcano: Planet X cataclysm imminent.

Chapter 12

1154. Major Planet X activity in California due to Planet X

Figure 1 below shows some photographs sent to me by Beth of the ground at a cemetery, which is not far from where she lives, but at a higher elevation, because it is on a hill. The amount of soil emerging from the ground at this location is even greater than at her own property, which has been discussed in previous articles (see Article 1140: Spherical objects and soil emerging from ground in California due to Planet X and Article 1147: Planet X effects accelerating: water emerges in California) [1, 2].

Figure 12.1. These clumps of soil seem to have fallen from above and would thus have formed on the surface of a Planet X core or Stellar Core (SC) in the sky, which is connecting to and absorbing energy from earth soil creating cores under the surface of the earth, which respond by also creating soil which then emerges from the ground.

Figure 12.2. Clumps of soil that have dropped from the sky and tiny sinkholes, which are similar to what was observed on Beth's property, although no cores are seen emerging from the ground. This does not mean that the earth cores under this hill are not moving toward the surface, but that they just have not yet emerged Sinkholes can also be seen.

Figure 12.3. Left: Earth soil creating core first observed emerging from the ground on Beth's property, in July, the soil it creates can be seen, on top of it, and further to the left. Some of this soil was most likely dropped by the SC in the sky connecting with it. **Right:** The same core in September: it has now emerged much further, from the ground, and a small core, most likely its baby, can now be seen also emerging to its left.

Figure 12.4. Sinkholes could also be seen in the newly emerging soil on Beth's property.

Figure 12.5. Lots of clumps of soil suggests that soil is both emerging and dropping from the sky, at the cemetery, and many holes (sinkholes) are forming as a result of the gravitational connections. Some may have said that this is due to gophers making holes in the ground but these do not suddenly appear in large numbers and the fact that the same type of activity was observed on Beth's property where cores were clearly observed emerging from the ground shows that this is due to Planet X.

Figure 12.6. Another photograph showing a clump, made up of mixture of brown and grey soil, which appears to have been dropped over this location. The grey soil seems to have been dropped on top of the brown soil suggesting that the SC concerned, first created brown and then grey soil. This may have been as a result of it connecting to a brown soil creating core first and then a grey soil creating core.

Figure 12.7. Another clump of only grey soil, this time, which has clearly been dropped from above at the cemetery, as it is on top of the vegetation.

Figure 12.8. Soil seems to be emerging close to this headstone, which also seems to be leaning suggesting that the ground movement as a result of new soil emerging has caused it to lean over. The soil seems to be below the vegetation and lifting it as it emerges from the ground.

In conclusion, soil seems to both be emerging from the ground and dropping from Planet X cores in the sky at a cemetery, which is not far from Beth's property, suggesting that this is something which is occurring over a very wide area, and possibly over the whole of California. This is a process that will eventually completely transform the ground and will thus damage houses. But also since earth cores, deeper within the earth, will also be affected, it is likely to lead to the formation of geysers, mud volcanoes and magma volcanoes throughout California as well as cause many, possibly catastrophic, earthquakes.

References:

[1] Albers, C. (2019). Article 1140: Spherical objects and soil emerging from ground in California due to Planet X.
[2] Albers, C. (2019). Article 1147: Planet X effects accelerating: water emerges in California

Chapter 13

1157. Small sinkholes at cemetery in California are not gopher holes

In Article 1154: Major Planet X activity in California due to Planet X [1], I talked about the soil that is both emerging from the ground and being dropped from the sky, due to Planet X cores making gravitational connections, at this location. The soil is emerging in a very similar manner to the way it is also emerging on Beth's property. But, the appearance of small holes, which I have explained are sinkholes and were also seen on Beth's property were attributed to gophers by many people and the huge quantity of new soil on the surface, to these rodents digging into the ground. However, the same piles have appeared on Beth's property suddenly. In order for gophers to dig up so much soil in so short a time there would have to be very many of them and they would have to suddenly have appeared, which is simply not possible. Gophers do not suddenly appear somewhere where they have never been seen before. In addition, the soil can often be seen to be emerging from the ground as the plants are clearly above the emerging soil. Gophers cannot cause the ground to bulge outwards with new soil.

Figure 13.1. Soil can be seen emerging from the ground and some have holes in them. Some of the soil appears to be dropped from above, but most appears to be emerging from the ground and it is this soil that has holes in it, at the cemetery. The holes would be sinkholes as they are the result of gravitational connections between Planet X cores and earth cores.

Figure 13.2. Soil emerging from the ground and small holes matching what is observed at the cemetery but this photograph is from Beth's property. If gophers were responsible for what happened at the cemetery, they would be responsible for what happened on Beth's property.

But how would gophers get spherical earth cores to emerge from the ground and cause soil to be dropped precisely on top of them, and how did they produce piles of soil in places where no holes are visible. Do Gophers also carry piles of dirt and place them strategically somewhere, far from holes they have dug up?

Figure 13.3. Earth soil creating cores emerge from the ground. Soil is seen emerging from the mound created by the core emerging on the right. No holes are seen here so gophers cannot possibly be responsible for this occurrence. Yet the soil is the same that is seen elsewhere where holes appear indicating that this is not the work of gophers but of Planet X.

Figure 13.4. A mound of light grey, almost white, soil, at the cemetery, is seen, indicating that a soil creating core, which creates this type of soil as an envelope, is emerging, at this location. More white soil seems to also be emerging in the foreground, but has not emerged to the same extent as the soil in the background. These are signs that cores are also emerging in the cemetery. But gophers cannot cause the ground to bulge outwards or move cores upwards, so this is not the work of gophers.

Figure 13.5. A tiny earth core with a grey almost white envelope, which completely emerged from the ground on Beth's property, is seen here. This core cannot in any way be the product of gophers at work and we can see from its envelope that it would create exactly the same type of soil that is observed all over Beth's property and the cemetery and that these are therefore the type of soil creating cores under the ground that Planet X is making connections with and causing to emerge.

Figure 13.6. Left: Holes (sinkholes) in Beth's property appear to be in soil that is emerging, as it is compact and below rooted vegetation, no piles of soil are seen to indicate that gophers were hard at work digging tunnels. **Right:** Soil which appears to also have emerged from the ground at the cemetery, as rooted plants can be seen above it, and large numbers of hole: Some of the soil seems to form small piles but not in an arrangement that would suggest that they had emerged from the holes due to gophers digging it out of the holes.

It thus likely that gophers have been used as the excuse of choice for the appearance of holes and piles of soil around the world, when in a lot of cases, the soil was simply emerging due to Planet X. The powers that be would use any excuse to cover up Planet X activity and have filled humanity's knowledge base with lies as a result. This is simply another of those lies to add to the list (see Article 1040: Aliens have filled humanity's knowledge base with lies and Article 1138: Planet X cover up: Aliens destroying humanity and Earth with lies) [2, 3].

Figure 13.7. A close up on some of the holes in the ground clearly shows plants which are rooted in the ground above the level of the holes, indicating that most of the soil is emerging from below. A small pile of soil to the left is not enough to justify the size of the holes. The soil which is usually seen on top seems to also be made of larger sized pieces than the soil which emerges from the ground, suggesting that the earth cores produce finer grained soil than the Planet X cores in the sky. The soil which seems to be dropped to the left of the 4 holes, seems to have been dropped in a curious ring pattern. You would not expect a gopher to be so neat that it would position dug up soil in a ring pattern.

Figure 13.8. Another patch of soil which seems to have emerged from the ground as rooted vegetation is seen on top of it, suggesting that the plants also moved upwards with the emerging soil. One of the holes in the ground is hexagonal, something which is sometimes seen in sinkholes but would in no way be expected from a gopher dug hole.

Figure 13.9. The formation of a sinkhole: The sinkhole forms under the surface, there is no direct force from the top, the ground under the surface moves downwards first, then the surface of the road falls inwards due to having lost the ground's support. This shows that the interaction leading to the formation is between the Planet X central core and the earth's central core, resulting in the earth moving the position of the earth's surface in response to the presence of the Planet X central core. A dome can be seen in the center of the hole, showing that this sinkhole formed as a result of a gravitational interference pattern with a central maximum (see Article 674: Sinkhole in Knoxville Tennessee due to Planet X and gravitational interference) [4]. The sinkhole has a hexagonal outer perimeter.

In conclusion, soil and cores are emerging at a nearby cemetery in a similar manner to the way the same is occurring on Beth's property. Holes can also be seen all over the emerging soil, which are clearly sinkholes. The idea that these could be produced by gophers is illogical and obviously another idea which has stemmed from efforts by the 'powers that be' to cover up Planet X effects. Gophers cannot suddenly appear out of nowhere in large numbers to dig holes in the ground, nor would they dig without leaving piles of soil close to the holes and instead pile the soil in neat ring patterns. Gophers cannot cause cores to emerge from the ground or to cause soil to bulge outwards underneath plants, nor can they create hexagonal sinkholes.

References:

[1]　Albers, C. (2019). Article 1154: Major Planet X activity in California due to Planet X.

[2]　Albers, C. (2019). Article 1040: Aliens have filled humanity's knowledge base with lies

[3]　Albers, C. (2019). Article 1138: Planet X cover up: Aliens destroying humanity and Earth with lies.

[4]　Albers, C. (2019). Article 674: Sinkhole in Knoxville Tennessee due to Planet X and gravitational interference.

Chapter 14

1172. Earth cores emerge from the ground in California due to Planet X

Beth has been sending me photographs of what has been occurring on her property since July, which seemed to have started with the appearance of huge Planet X cores in the skies above her location, in northern California. One of the first photographs she sent me was of an earth soil creating core, which had started emerging from the ground. This observation allowed me to understand that Planet X connections cause earth cores to emerge from the ground and when these cores are much larger magma creating cores, it leads to the formation of volcanoes. The same process leads to earthquakes, i.e. cores, within the earth, respond to connections with Planet X cores, by ejecting matter and moving toward the surface of the earth, which results in earthquakes.

Figure 14.1. Left: Earth soil creating core first observed emerging from the ground on Beth's property, in July, the soil it creates can be seen, on top of it, and further to the left. Some of this soil was most likely dropped by the SC in the sky connecting with it. **Right:** The same core in September: it has now emerged much further, from the ground, and a small core, most likely its baby, can now be seen also emerging to its left.

Figure 14.2. Sinkholes in the newly created soil on Beth's property.

Earth soil creating cores all over Beth's property have been activated by Planet X and have as result been creating and ejecting new soil, which appears on the surface. In addition, her property is covered in small sinkholes. These sinkholes have led to the ground caving in, in places (see Article 1168: Planetary wide Planet X cataclysm starting in California) [1]. The fact that sinkholes as well as mounds occur, suggests that the earth cores respond in two different ways to Planet X, one way leads to new material being ejected toward the surface, which leads to the appearance of new soil, and the other leads to existing soil retracting backwards and into the earth core, which results in sinkholes.

Figure 14.3. The same earth core, which was first observed emerging from the ground, with the smaller companion or baby, next to it, on October 3rd 2019. The earth cores are progressively emerging and at the same time they are creating soil, which causes a larger and larger mound of earth to form, at this location. Increasing amounts of new soil can be seen above the two cores, suggesting that gravitational connections with huge Planet X cores in the sky are occurring on a continual basis. New soil most likely ejected by a third earth soil creating core seems to be emerging to the right of the first core to start emerging.

Figure 14.4. Another photograph sent in by Beth of the clumps of soil which she has pulled out of the ground and which have tiny holes on the bottom where they were connected to the soil creating core responsible for their emergence.

Figure 14.5. Several holes can be seen on the bottom of this chunk of newly created soil, suggesting that the soil creating core responsible for the appearance of this soil made several ejections from different points on its surface. This material also seems to be denser than the soil that was first seen emerging suggesting that cores from deeper within the earth are responsible for it.

The connections, which cause soil to appear on the ground are somehow reminiscent of trees growing, branches sprouting and reaching out, and indicate that the same patterns appear at every level in the universe, to the point that planets grow in a similar manner to trees. The changes are natural but they are profound and will be happening at all levels, i.e. also deep within the earth, as evidenced by the number of earthquakes that continuously take place. What we are observing in California is surface reformation, which will ultimately change the ground to such an extent that it will most likely destroy most buildings and is likely to also lead to new volcanoes forming and large portions of land sinking, as well as to strong earthquakes. It is thus a truly cataclysmic event, which is unfolding in California and will ultimately affect the whole planet.

Figure 14.6. Cylindrical cloud in the sky: A cloud can only maintain this shape if it contains solid matter within it. This is therefore a connection like the ones that earth cores are making in order to create soil on the surface, but which was done inside one of the Planet X planets, and since it is so much larger, it must have been produced by a much larger core within the planet (**Source**: MrMBB333 video, July 6th 2019).

In conclusion, large scale surface reformation continues to occur in California and although the process is fascinating to see, this is not a benign process, as it must be similarly affecting cores that are much larger and much deeper within the earth, so that this is an unfolding cataclysmic event.

References:

[1]   Albers, C. (2019). Article 1168: Planetary wide Planet X cataclysm starting in California

Chapter 15

1176. Earth cores emerging and sinkholes forming in California due to Planet X

Figure 1 below shows one of the recent photographs sent in by Beth of her property, where since July of 2019, earth soil creating cores and an untold number of small sinkholes have appeared. The earth cores are progressively emerging and producing larger and larger mounds on the surface. New emerging soil can be seen emerging from the ground, every few feet, and many small sinkholes have formed and continue to from, in the newly created soil, so that Beth can see changes and increasing numbers of sinkholes every single day (see Article 1172: Earth cores emerge from the ground in California due to Planet X) [1].

Figure 15.1. A photograph of the first soil creating core to be observed emerging from the ground on Beth's property, taken from a distance in order to provide perspective on the size of the object: The object is clearly not alone as two other cores seem to be emerging in very close proximity to it.

Figure 15.2. Another photograph provided by Beth of the same object, where we can see 2 smaller mounds forming next to the larger mound. Beth has added a 24 inch T square to offer perspective, as to the size of the core and the mound it is creating.

These photographs provide a view into the creative processes that lead to the formation of planets as well as the processes that will be going on at much greater depths with earth cores, which will be much larger and much deeper within the body of the earth that will be responding to Planet X gravitational connections, in the same way. Some will create matter and move closer to the soil and some will be creating matter and also responding by reabsorbing some of the matter that they have created thus giving rise to sinkholes. When these cores are much larger and much deeper within the earth what that gives rise to is earthquakes.

It is obvious that a large number of earth cores have been activated on Beth's property because new mounds of newly emerging soil can be seen every few feet and many of them have large number of sinkholes as well. Since the ground appears to be caving in, at many locations, within the property, it is likely that sinkholes can form which are deeper within the earth and may not go all the way to the surface. This then suggests how this process in the case of larger cores, within the earth, creating sinkholes deep underground may cause large regions of the surface of the earth to sink.

Figure 15.3. Some of the sinkholes that keep appearing on Beth's property, with her shoes (size 8.5) in the picture, to allow perspective as to their size: The sinkholes seem to be between 2 and 3 inches in diameter.

Figure 15.4. The sinkhole has very clean vertical sides as if a precision instrument cut the soil away and removed it. However no metallic or laser cutting tool would be able to do this as the roots from plants have not been touched. The soil inside the cylindrical volume just vanished.

Even an animal digging a hole would have done more damage to the plant roots than seems to have occurred here. These cannot in any way be done by any animal. And the animal would have left a mound of dug up soil as well, which clearly is not present next to any of the sinkholes (see Article 1157: Small sinkholes at cemetery in California are not gopher holes) [2]. The soil in the sinkhole simply vanished.

Figure 15.5. A view inside a sinkhole shows that it has clean vertical sides as if the ground that was once in it suddenly disappeared. This shows that cores are able to create matter and cause it to emerge, as well as cause the matter they have created, to disappear or retract, within a gravitational connection region.

In conclusion, earth soil creating cores all over a property in northern California have been activated by huge Planet X cores, in the sky, which are obviously making gravitational connections with these earth cores and causing them to emerge from the ground, create matter as well as create sinkholes. This is a process that will also be happening deep underground, with much larger earth cores, and thus lead to earthquakes and the appearance of volcanoes, and also explains how large regions of the surface can suddenly sink after an earthquake, i.e. a core, deep within the earth, retracted some of its created material, thus creating a sinkhole above itself, deep within the ground and far from the surface of the earth but which resulted in the ground above, to drop down into the newly created empty space.

References:

[1]   Albers, C. (2019). Article 1172: Earth cores emerge from the ground in California due to Planet X.
[2]   Albers, C. (2019). Article 1157: Small sinkholes at cemetery in California are not gopher holes.

Chapter 16

1181. Planet X depleting California cores: darkness and light sinkholes

Figure 1 below shows a photograph sent in by Beth of the sky in California close to sunset and it is very red. This is sunset, 6:50 pm, local time, on October 6th 2019, and the camera was facing eastwards and thus away from the Sun. It shows that the lower atmosphere is emitting red light rather than blue. Since red light has a lower frequency than blue light and thus is lower in energy than blue light, this is an indication that earth's light emission cores are becoming energy depleted.

Figure 16.1. Photograph sent in by Beth of the sky over her location in California facing eastwards at sunset. The red color in the atmosphere close to the horizon shows that the earth's atmosphere is emitting red rather than the normal blue.

Figure 16.2. The sky over London from August 24 2019, as seen from a plane on approach to London, was also red close to the horizon (see Article 1079: Earth's atmosphere emits red light close to the surface of the earth) [1].

Daylight and the earth's blue sky is due to the earth's atmosphere emitting light (see Article 888: The Sun is gone: Daylight comes from earth's core and Article 1059: Blue sky after sunset indicates the presence of Planet X in earth's skies) [2, 3].

Figure 16.3. The sky over Syria close to sunset on September 3rd 2019 showing a red sky just above the earth's surface, indicating that earth's lower atmosphere is emitting mainly red light at lower altitudes close to sunset, a sign of the core system's energy depletion (**Source:** Albin Szakola via Twitter).

The earth's atmosphere close to the horizon over Syria also has a hint of red, orange and pink, but not as intense as over London and even less intense than over California, which suggests that the cores under California are at a more advanced stage of energy depletion and they seem to have been for at least a year now.

Figure 16.4. Photograph of the sky over Beth's property on October 7th 2018, at 6:45 pm, about a year before the photograph shown in figure 1 was taken, and only from 5 minutes before.

An advanced stage of energy depletion is likely to lead to increased Planet X effects such as the collapse of hillsides and the opening of fissures in the ground, due to the ground losing cohesion. It also indicates that Planet X is making more connections over California than anywhere else and thus explains why there is so much activity from earth soil creating cores and earthquakes (see Article 1176: Earth cores emerging and sinkholes forming in California due to Planet X and Article 1168: Planetary wide Planet X cataclysm starting in California) [4, 5].

The Sun stopped emitting visible light as a result of the same process which we see occurring here. The fact that some regions are more depleted than others also suggests that some regions can become darker than others first. This is a process that will eventually make the earth's atmosphere dark all the time but there can also be temporary darkness, as has occurred in Russia now twice and once in Argentina, where the earth's light creating cores stop sending out energy through the atmosphere to cause it to emit light (see Article 1056: Darkness instead of daylight again in Siberia due to Planet X) [6]. This can be described as a light sinkhole; the cores pull the energy back into themselves, just like they pull matter back into themselves causing a portion of ground to disappear thus creating a sinkhole. These are due to gravitational connections between the earth's core system and the energy depleted Planet X cores.

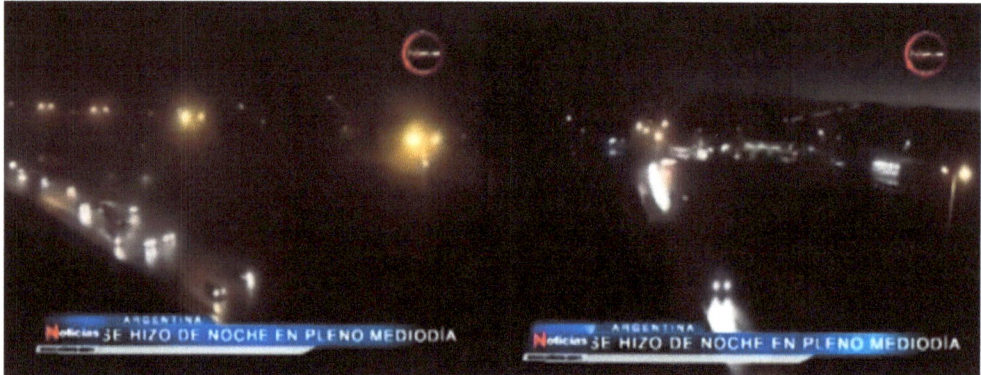

Figure 16.5. Screenshot from a television news report in Argentina on August 12th 2011, where the sky went completely dark in the middle of the day: An extremely large region went completely dark as we can see just a little greyish light close to the horizon in the distance in the screenshot on the right.

The above event would be due to the light producing cores in the body of the earth suddenly stopping the flow of energy through the atmosphere due to the approach of Planet X and due to making gravitational connections with it. Instead of the usual outflow of energy, these earth cores pulled the energy inwards back to themselves, thus producing a light sinkhole in the atmosphere.

Figure 16.6. Some more screenshots, which show the extent of the darkened region, as grey light can be seen close to the horizon, in the distance, as the camera points in different directions. The light close to the surface of the earth also shows where daylight in the earth's atmosphere comes from. The illumination and thus daylight originates with the earth's core not the Sun, although the Sun is supposed to add to it, which is why it is lighter close to the surface than in the sky above.

Notice that this could not have been due to a cloud moving over the region or an eclipse because no cloud or eclipse causes sudden pitch darkness, as if the lights had been turned off, but just like sinkholes can suddenly form, due to earth cores retracting matter, in response to Planet X, a light sinkhole can suddenly occur for the same reason.

In conclusion, red light being emitted by the atmosphere in California suggests that the light producing earth cores are more energy depleted in this region of the world, than at other regions. Temporary darkness as has occurred over in Argentina and Russia can be described in terms of a light sinkhole and is due to light creating earth cores retracting the energy that allows the atmosphere to emit light.

References:

[1]   Albers, C. (2019). Article 1079: Earth's atmosphere emits red light close to the surface of the earth.
[2]   Albers, C. (2019). Article 888: The Sun is gone: Daylight comes from earth's core.
[3]   Albers, C. (2019). Article 1059: Blue sky after sunset indicates the presence of Planet X in earth's skies.
[4]   Albers, C. (2019). Article 1176: Earth cores emerging and sinkholes forming in California due to Planet X.
[5]   Albers, C. (2019). Article 1168: Planetary wide Planet X cataclysm starting in California.
[6]   Albers, C. (2019). Article 1056: Darkness instead of daylight again in Siberia due to Planet X.

Chapter 17

1185. Planet X activity in California and the amazing universe we live in

Figure 1 below shows a photograph sent in by Beth, which is absolutely astounding. It shows a piece of dense soil in the shape of a mushroom. If any photograph could ever show how the earth is not made out of inert soil, this is it. The only way that such a formation is possible, is if the soil grew out of the ground and is still connected to the core that created it, because a rock of this shape would not be stable and be able to remain upright, it would have fallen over. So this photograph is amazing evidence attesting to the fact that planets are created by cores. It also agrees with the conclusion I came to in Article 1184: Static electricity on human body generated by the human creative spirit [1], i.e. that the universe we are living in, is a spiritual universe.

Figure 17.1. A piece of soil juts out of the ground, it is in the shape of a mushroom and would not be stable at all and would have fallen over, if it was not attached to the core that had created it, thus showing that the surface of the planet and that a planet is made of is created by core matter.

I have thought for a very long time that we are living in a physical universe, which is imbedded in a spiritual universe, with a larger number of dimensions. But what I have found that this is not true. We are living in a universe made of light, and light is spiritual energy, so we are living in a spiritual universe, there is no other universe that this one is imbedded in [1]. For some reason there are things in the universe that we do not see, due most likely to what happened to us at the Garden of Eden, which changed the human body to being able to detect only visible light, but this is the only universe there is. So the idea that there are other universes with more dimensions is another of the Serpent's lies (see Article 1040: Aliens have filled humanity's knowledge base with lies) [2].

Figure 2 below is a perfect demonstration of what I stated in previous articles and that is that cores can retract matter, which creates sinkholes, but sometimes the sinkhole does not reach all the way to the surface. However, the sinkhole's presence below the surface can cause the ground above it, to collapse into it, thus causing a hole as shown below.

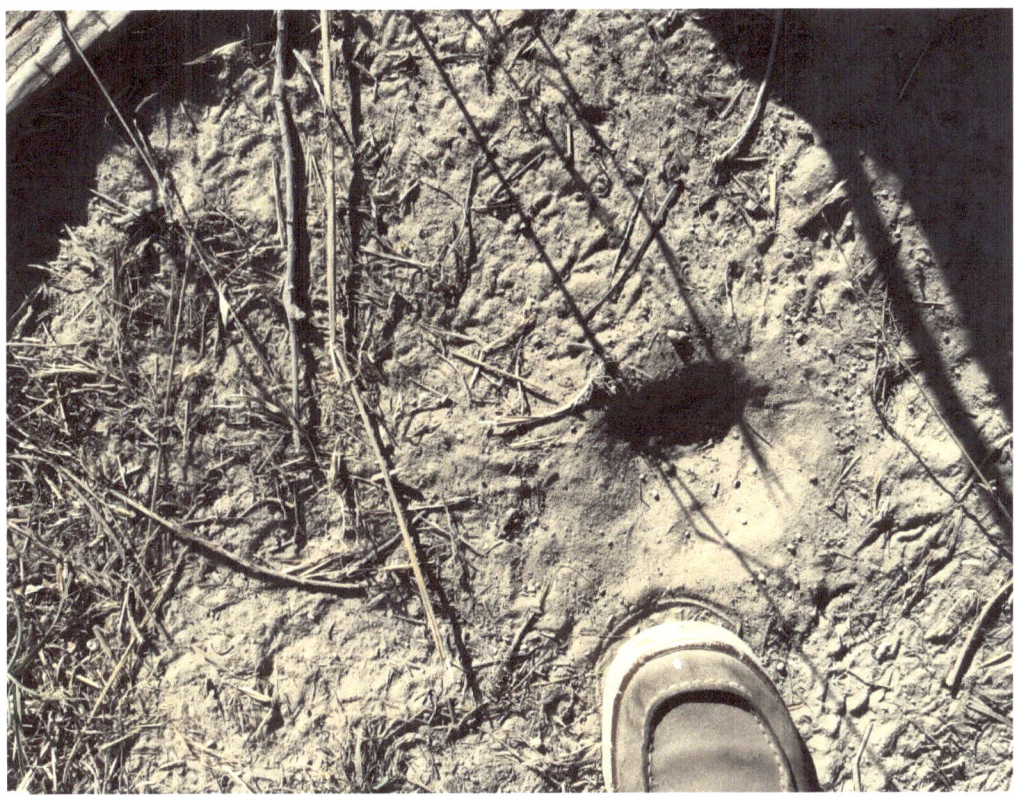

Figure 17.2. This hole does not have perfectly cut sides but seems to be made of soil dropped into a hole below the surface and is thus an example of surface soil dropping into a sinkhole that formed below the surface.

Figure 17.3. Beth finds new pieces of emerging soil every day; these are now occurring every few feet and show that Planet X is causing the surface to reform at this location, a process that is far from benign as the same processes are sure to occur much further underground, with much larger cores as well.

In conclusion, we are living in an amazing universe, where planets are created by cores, which use light as energy and where everything is made out of light, and where soil can be in the shape of a mushroom. This is also the only universe there is, this is the spiritual or light universe that God created for us.

References:

[1] Albers, C. (2019). Article 1184: Static electricity on human body generated by the human creative spirit.
[2] Albers, C. (2019). Article 1040: Aliens have filled humanity's knowledge base with lies.

Chapter 18

1193. Roll clouds: Planet X core systems broke apart

Figure 1 below shows a roll cloud in earth's skies. These are long cylindrical cloud formations, which cannot possibly maintain their shape unless there is solid matter in them, i.e. Planet X core matter. This roll cloud appears to be at least tens of miles in length. But from studying earth cores emerging from the soil at one property, in California, thanks to Beth who has been sending me the photographic evidence, it can be understood what these long pieces of Planet X core matter are.

Figure 18.1. Long roll cloud hovering in earth's atmosphere.

Figure 18.2. Left: Piece of material in the shape of mushrooms which juts out of the ground: It is what remained when the core which emerged from the ground in Beth's property, shown on the right, was stepped on. **Right:** The earth soil creating core which originally emerged from the ground.

Figure 18.3. This piece of ground clearly emerges from the ground, it appears to be cylindrical in shape at the base and indicates that the core that emerged was connected through a cylindrical shaped piece of ground to another core which must have been deeper underground.

Figure 18.4. Left: Photograph from October 11th 2019 of another soil creating core, which emerged from the ground on Beth's property. **Right:** When broken into pieces, the core had one central cylindrical piece that must have been the piece from which the core originated after its parent core sent out the connection, showing that core systems seem to grow like root systems in the plant kingdom.

Thus, core systems seem to be connected by cylindrical pieces of core matter and this is what roll clouds are; they are the long pieces of core matter connecting the different cores making up the Planet X planets' and stars' core systems. The fact that they are no longer connected shows that the core systems broke apart when the Planet X planets and stars broke apart.

Figure 18.5. The end of a roll cloud which is not connected to a core clearly shows that the core that was connected to the end of the cylindrical piece of core matter broke off.

Figure 18.6. Core matter connections are not always perfectly cylindrical or even perfectly straight; this one has a bend in it and is thicker on the bottom. These very long connections would most likely occur either close to the deepest part of a planet or inside a star core system (see Article 956: Roll clouds have Planet X core matter in them) [1].

Figure 18.7. Craters on the moon: The moon is a Planet X planetary central core (see Article 1058: The Moon is a Planet X planetary central core and not a spaceship) [2] and thus the craters on its surface are the points from which the cylindrical connections emerged and went out to form satellite cores, which would in turn produce their own core sprout.

Thus craters on the surface of the moon would have been connection points or points from which energy would flow out from the central core to the satellite cores, it had created, which would essentially be its baby cores.

In conclusion, roll clouds inside earth's atmosphere, which must have solid core matter within them, must be cylindrical connections between cores that have also been observed in the earth's own core system, as new earth soil creating cores emerge from the ground due to Planet X. Roll clouds are also evidence to the fact that the Planet X core systems broke apart, the cylindrical connections breaking off the spout from which they first emerged on the originating core, leaving a crater behind.

References:

[1] Albers, C. (2019). Article 956: Roll clouds have Planet X core matter in them.

[2] Albers, C. (2019). Article 1058: The Moon is a Planet X planetary central core and not a spaceship.

Chapter 19

1195. Sinkholes: caused by Planet X and show that the universe is alive

Sinkholes are holes in the ground, which seem to occur for no apparent reason. They come in many different sizes as impact craters are actually sinkholes. The observation of tiny sinkholes forming on a property in California as earth soil creating cores emerged from the ground brought increased understanding of what sinkholes are and what causes them. Figure 1 below shows the formation of a sinkhole.

Figure 19.1. The sinkhole forms under the surface, there is no direct force from the top, the ground under the surface seems to move downwards first, then, the surface of the road falls inwards due to having lost the ground's support.

Figure 19.2. A sinkhole on Beth's property where a large number of sinkholes have formed at the same time that new soil creating earth cores are emerging from the ground. The sinkhole has very clean vertical sides as if a precision instrument cut the soil away and removed it. However no metallic or laser cutting tool would be able to do this as the roots from plants have not been touched. The soil inside the cylindrical volume just vanished and shows that the core responsible for the soil in which the sinkhole formed retracted the material it had created, it reabsorbed it (see Article 1176: Earth cores emerging and sinkholes forming in California due to Planet X) [1].

All matter in the universe is made out of photons and thus energy, hence the absorption of matter is the same as energy reabsorption and seems to naturally occur whenever a core is about to eject core matter, which starts with a long cylindrical connection and then allows the formation of a new baby core at the opposite end of the cylindrical core connection. Roll clouds in the earth's atmosphere are connections between parent and child cores and a part of the Planet X core systems, which broke apart, thus destroying the Planet X planets and stars.

Figure 19.3. Left: The central piece of a core which emerged from the ground on Beth's property, which remains upright due to the cylindrical connection which disappears into the ground. **Right:** Roll cloud: a Planet X core matter connection (see Article 1193: Roll clouds: Planet X core systems broke apart) [2].

When the Planet X planets and stars broke apart the connections broke off the points where they had been connected, leaving craters behind, which shows that the object creating the connection first retracted some matter, at the point where the connection was to be made, and then ejected matter out through it. Thus, sinkhole formation is a natural process in the growth of core systems in the universe.

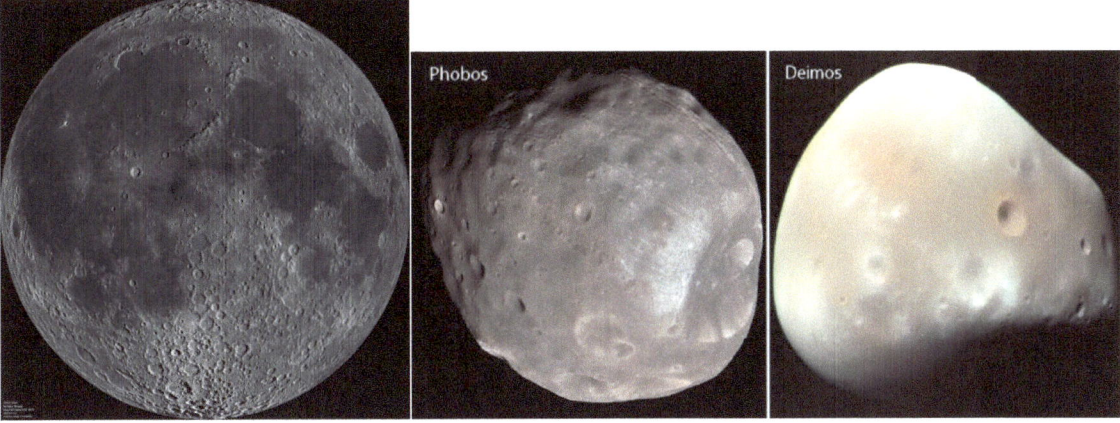

Figure 19.4. **Left:** Craters on the moon: The moon is a Planet X planetary central core (see Article 1058: The Moon is a Planet X planetary central core and not a spaceship) [3] and thus the craters on its surface are the points from which the cylindrical connections emerged and went out to form satellite cores, which would in turn produce their own core sprout. **Center and right:** Mars' moons: Phobos and Deimos, are very small only 6 and 4 miles in radius, respectively, but they like earth's moon are Planet X cores and have craters or sinkholes on them, which would have been connection points from which they ejected matter by first retracting some matter.

Figure 19.5. Large meteor crater in Arizona, it appears to be perfectly circular and to have a dome structure in the center. Perfectly symmetric craters with equally sloped sides all around cannot form due to an impact, they have to be sinkholes, i.e. they had to form as a result of earth's core system retracting material in order to follow with ejection or outflow of energy in response to a Planet X core above that point, on the surface of the earth (see Article 675: Sinkholes all over the earth: impact craters are sinkholes) [4].

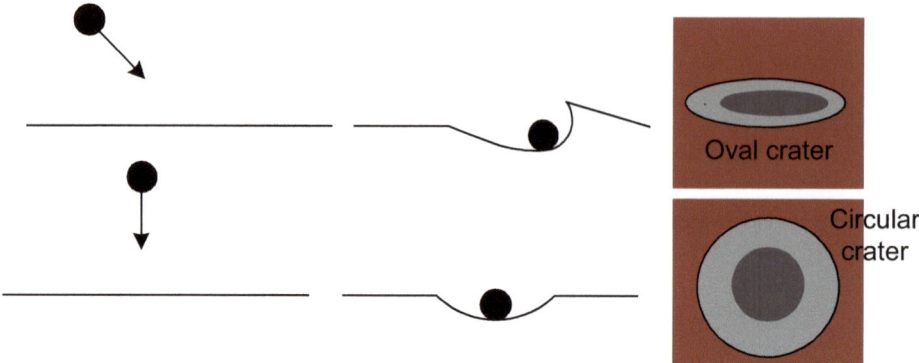

Figure 19.6. An impact crater cannot be expected to be circular in shape unless the impacting object comes in at an angle of 90°. But all objects in the universe follow curved trajectories around a massive object and will thus never come in at right angles, making a circular impact crater impossible. A real impact crater will always be oval and deeper on one side. Thus, circular craters are not impact craters, they are sinkholes (see Article 675: Sinkholes all over the earth: impact craters are sinkholes) [4].

In the case of these huge sinkholes, the ground also rises at the outer edge of the sinkhole, so it is not just a matter of retracting but forming a sort of knot on the surface of the earth somewhat like what happens with trees.

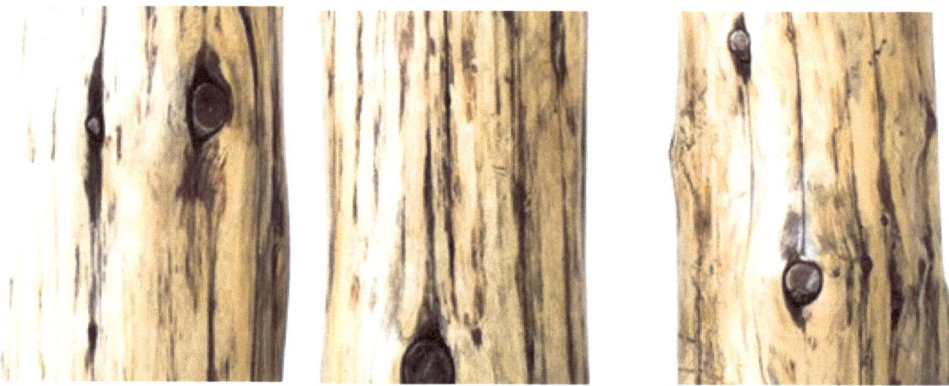

Figure 19.7. A tree's trunk opens up in what is called a knot from where a branch then grows outwards, sinkholes are like these knots and appear in response to gravitational connections between Planet X cores and the earth's core system. Notice that the branches do not just push through the outer trunk layer, the trunk opens to accommodate the new branch growth and in the same way, sinkholes form in order to accommodate an outward directed flow of energy.

Thus, craters on the surface of the moon and other Planet X cores formed when they were inside the Planet X planets and stars as they formed and grew. But craters on the surface of the planets in the Solar System formed in response to the presence of Planet X in the Solar System. As Planet X made gravitational connections with solar system core systems, their energy depleted status triggered an energy outflow process which is supposed to start with retracting of material thus causing the appearance of sinkholes. But these sinkholes can happen on the surface of the earth and deeper underground and must thus depend on how far the ejecting core's gravitational field extends, if it extends all the way to the surface the sinkhole appears on the surface, so we would expect a connection with the earth's central core to create a huge sinkhole on the surface of the planet as the surface of the planet is an extension of the central core's own surface. But smaller cores below the surface may create sinkholes that never reach the surface.

Figure 19.8. This hole does not have perfectly cut sides but seems to be made of soil dropped into a hole below the surface and is thus an example of surface soil dropping into a sinkhole that formed below the surface.

In addition, craters which form at the mouth of volcanoes are also sinkholes as the magma creating core inside the earth's core system retracts material and thus creates the sinkholes before it ejects matter, i.e. magma. The magma will reach the surface if its field extends all the way to the surface and thus the sinkhole will form on the surface. But a magma creating core's gravitational field does not extend all the way to the surface it will create a sinkhole and eject matter below ground which will only be detected as an earthquake.

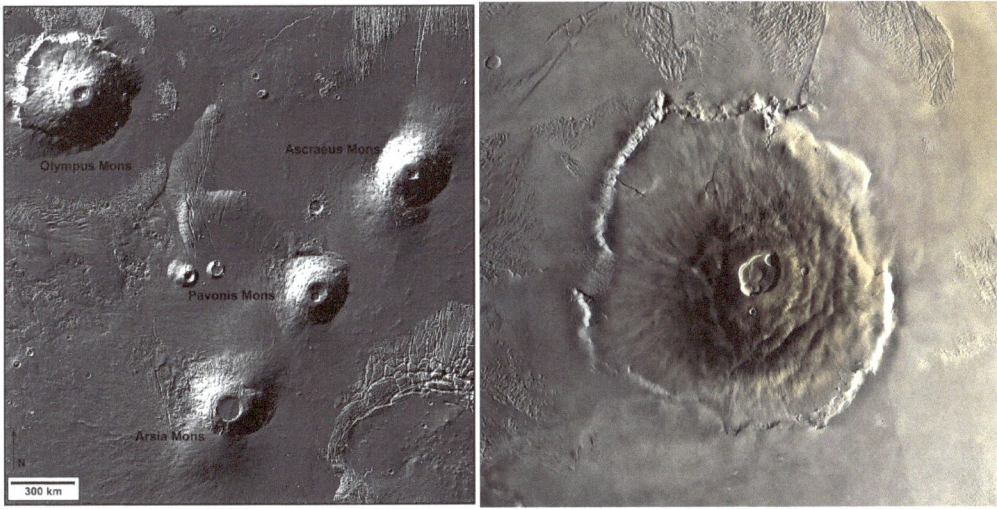

Figure 19.9. Olympus Mons is the largest volcano on Mars and in the Solar System as it is 22 km (12.4 miles) high. The other three are also huge, over 10 km in height. Olympus Mons is 640 km (397 miles) wide. They are topped by sinkholes, as in the case of volcanoes, no magma would reach the surface if the core was not able to retract matter from the surface at the beginning of the ejection process, which occurs in response to a gravitational connection with Planet X (see Article 1082: Planet X effects on Mars: volcanoes and craters) [5].

Gravitational connections between Planet X cores and earth cores allow Planet X cores to create the matter that the earth core they are connecting with creates and they then drop it in the connection region. This is an indication that energy is flowing to it from an earth core that is under the surface of the earth that would be doing some of its own ejection of material whilst at the same time retracting material and creating sinkholes which may be deep underground and thus not visible on the surface. But, the appearance of a sinkhole on the surface is a sign that energy is flowing through the atmosphere from that point up to a Planet X core in the sky.

Figure 19.10. Gravitational connection between Planet X cores in the sky and the earth's core system leads to connections made of material which the Planet X cores create, this could be fine soil or dust or water vapor, and sometimes droplets of water, i.e. rain drops. Energy flows through the connection and the denser the material in it the faster the energy transfer to the Planet X core. A sinkhole may not be apparent on the surface at these locations, but sinkholes and matter ejection is most likely happening deep underground above the core that the Planet X core is connecting gravitationally to here (**Source**: YouTube Video by The Two Preachers, 11 October 2019) [6].

The size of a sinkhole would thus be relative to the amount of energy that is to be transferred and that depends on the size of the earth core transferring energy and the size of the connection point on the surface of the receiving Planet X core.

In conclusion, sinkhole formation is a part, of the natural energy transfer mechanism, which leads to cores, ejecting core matter, and thus growing or procreating. Thus, sinkholes on Planet X cores which have been adopted by Solar System core systems and are now in orbit as satellites around these objects have craters or sinkholes, which formed due to this natural process, whilst sinkholes on the surface of the Solar System planets form as a result of gravitational connections between Planet X cores and the earth's core system. Sinkholes are like knots in trees and thus add to the evidence indicating that the universe is alive.

References:

[1] Albers, C. (2019). Article 1176: Earth cores emerging and sinkholes forming in California due to Planet X.
[2] Albers, C. (2019). Article 1193: Roll clouds: Planet X core systems broke apart.
[3] Albers, C. (2019). Article 1058: The Moon is a Planet X planetary central core and not a spaceship.
[4] Albers, C. (2019). Article 675: Sinkholes all over the earth: impact craters are sinkholes.
[5] Albers, C. (2019). Article 1082: Planet X effects on Mars: volcanoes and craters.
[6] Video by The Two Preachers, 11 October 2019.
https://www.youtube.com/watch?v=alrAu3PleIk&list=LLURC-iCzVpz1Jl010tIaZ8Q

Chapter 20

1168. Planetary wide Planet X cataclysm starting in California

Beth, has since July 2019, been sending me photographs of the changes that have been occurring on her property, which have indicated that earth cores emerge from the ground, as a result of gravitational connections with Planet X cores, in the sky, which have come in to the earth in order to absorb energy from the earth's core system. Her property is also full of tiny sinkholes, which also form as a result of the gravitational connections (see Article 1154: Major Planet X activity in California due to Planet X) [1]. She has now gone on a walk, which allowed her to see what the ground, outside of her property, looks like, and she has found signs of the same type of activity everywhere. Basically Planet X is activating earth cores in this region, and they are reforming the earth's surface in response.

Figure 20.1. Blobs of new soil appear everywhere. The blobs on the right are darker than the ones on the left. However, these blobs seem to be a lot denser than the ones I observed in earlier photographs of the soil emerging from the ground on Beth's property. This would suggest that this soil comes from connections with earth soil creating cores that are larger and deeper within the earth.

Figure 20.2. Another blob of newly created soil on the ground: It looks like it was deposited on top of the grass and thus that it dropped from above. But the next photograph refutes that.

Figure 20.3. Blobs of soil, which Beth pulled out of the ground, and seemed to be connected to something below it: It has holes on the bottom which suggests that this material was ejected by an earth soil creating core, so that the blob emerged onto the surface of the ground. This suggests that a long cylindrical connection was present between this material and the material on the ground.

Figure 20.4. Roll clouds in the sky suggest that these are the same type of connections as what is observed with the soil creating cores, except roll clouds are at a far larger scale and due to much larger cores than earth soil creating cores ejecting matter (see Article 956: Roll clouds have Planet X core matter in them) [2].

Figure 20.5. Two soil creating cores, side by side, seem to be emerging and creating soil.

Figure 20.6. Sinkholes in the newly created soil.

Beth has reported today, October 1st 2019, that after a walk around her property that the sinkholes have greatly increased in number and that the soil seems to be literally caving in in many places due to the sinkholes. This suggests that the surface reformation taking place in this region of the world and most likely through the whole of California, evidenced by the large number of earthquakes that seems to happen there every day, is accelerating and that the surface reformation can cause the ground to lift up and the ground to also sink.

Figure 20.7. Large number of earthquakes, in California, indicate that a large number of earth cores deep within the earth are being activated and thus ejecting matter or retracting matter and thus causing large scale reformation in the ground under California: There is also one earthquake in Oregon and one in Washington on this map as well as a 3.1 in Nevada and even one in Oklahoma, suggesting cores are also being activated in these regions and that therefore surface reformation is also occurring there.

Figure 20.8. Left: The first earth soil creating core observed emerging from the ground on Beth's property **Right:** A new mound covered in new emerging soil indicates that another core is emerging at this location, also on Beth's property (see Article 1147: Planet X effects accelerating: water emerges in California) [3].

With such intense activity occurring where Beth lives, which is North of San Francisco; it is actually surprising that there are not even more earthquakes occurring. However, it is likely that most of the earthquakes are not being reported. Such large scale surface reformation is going to affect entire cities in these regions and ultimately the whole world, so this is truly a planetary wide cataclysm, in the making.

In conclusion, observing accelerating reformation processes due to Planet X with the help of Beth, in California, suggests that what is occurring is truly a planetary wide cataclysm that is likely to be felt first in California.

References:

[1]   Albers, C. (2019). Article 1154: Major Planet X activity in California due to Planet X.
[2]   Albers, C. (2019). Article 956: Roll clouds have Planet X core matter in them.
[3]   Albers, C. (2019). Article 1147: Planet X effects accelerating: water emerges in California.

Chapter 21

1141. Asteroids coming in due to Planet X: Earth being destroyed

Asteroids are jagged pieces of rock, which come in from space. As I have shown in many previous articles, planets form from the inside out, the body of a planet is created by a core system and it forms a uniform solid surface. Soil creating cores close to the surface tend to form small particles of soil and solid rock formations appear to be due to magma solidifying, as a result of volcanic eruptions, which although due to Planet X coming in to absorb energy still solidify into uniform rocky ground. There is no reason to expect broken pieces of rock to appear on the ground. Yet, broken pieces of rock, i.e. rocks with jagged edges, are found all over the surface of the earth and often buried in a few feet of soil. Figure 1 below shows one of these broken pieces of rock, sent in by Beth, who has also sent me photographs of the soil creating cores, which are emerging from the ground on her property (see Article 1140: Spherical objects and soil emerging from ground in California due to Planet X) [1]. This particular rock appeared above one of the growing mounds, which she has been observing and is thus a new arrival. Some of the new soil, which is emerging from the ground, which the soil creating cores create as a result of the Planet X gravitational connections, in the same way that magma creating cores create magma in response to the same connections and thus give rise to volcanic eruptions, can be seen below the rock.

Figure 21.1. Photograph of a rock sent in by Beth who found it on her property above one of the emerging mounds, which are due to earth cores emerging from the ground as a result of Planet X gravitational connections.

The Planet X planets broke into pieces due to energy depletion, thus turning into a jumbled mess of satellite cores and a debris field made up of the broken pieces of planet or broken pieces of rock.

Therefore, Planet X systems coming into the Solar System are the reason why broken pieces of rock, i.e. asteroids are found all over the Solar System and the reason why these rocks are coming into the earth's atmosphere and finding their way to the surface.

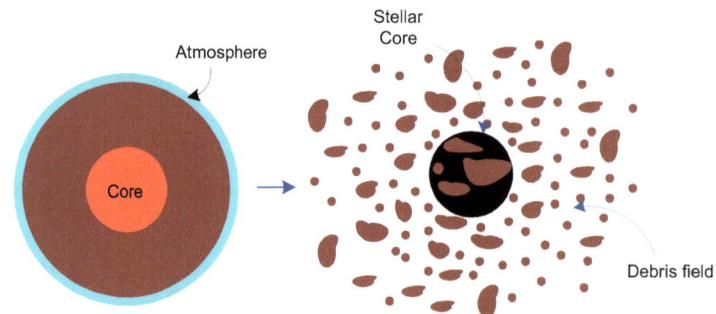

Figure 21.2. The Planet X planets broke up, the body of the planet turning into a debris field of broken pieces of rock and thus into an asteroid field.

Figure 21.3. The rock appears to be about 15 inches in diameter. But these asteroids or broken pieces of planet come in many different sizes.

Since they are all depleted in energy, when they first come into the earth's atmosphere, they will absorb energy as they sink through the atmosphere, the smaller rocks will absorb energy fast and will be able to fall through the atmosphere quite fast and often come down like hail, which is also a part of the Planet X debris fields, but larger pieces absorb energy slower and thus sink through the earth's atmosphere at a much slower speed and will eventually settle on the ground quite gently.

Figure 21.4. A huge granite rock by the name of Giant Rock lies in the desert floor in California. The rock is completely different from all the local rock formations and thus could only have come from above, i.e. it is a piece of one of the broken Planet X planets or an asteroid (see Article 983: Giant Rock in the desert: it came from outer space) [2]. Notice that there is no sign of any crater due to it having impacted the surface. This is because it is a huge piece of rock and would thus come down slowly and settle on the ground gently.

Figure 21.5. A huge rock appears amongst the clouds in Peru: this huge rock is suspended in the sky as it has not yet gained enough gravitational energy to have reached the surface of the planet. It will sink down slowly, the fact that it is long and flat indicates that it is a surface piece as the pieces that come in with buildings on them seem to be long and flat indicating that the surface of the planets broke into long flat pieces of rock (see Article 1134: Intelligent life all over the Universe) [3].

Figure 21.6. Left: The stones falling on the cars are brown but there seems to be quite a few white stones on the ground, which are most likely hailstones, which shows that both hailstones and rock stones fell from the sky. Also the raindrops on the window suggest that this occurred during a rainstorm. **Right:** Stones raining down from the sky in Turkey in 2017 (see Article 917: Stones fall from the sky in Romania: Planet X debris) [4].

Figure 21.7. Another photograph sent in by Beth of the rock she found on her property. The black patches are likely be due to an electrical interaction, which often occurs when meteors come into the earth's atmosphere. Meteors are also Planet X debris but they seem to be better drawers of energy which manifests as an electric current, most likely due to having a higher metal composition. Rocks that produce meteor type effects are usually smaller rocks and thus most likely from deep inside the planet, as the surface seemed to break into large pieces and the inside part into smaller pieces of rock.

Figure 21.8. Various photographs showing pieces of rock which fell from the sky over Cuba on February 1st 2019. The rocks are black on the outside due to the chemical reactions, which resulted from a larger electron current, flowing into it (see Article 779: Meteors are Planet X debris: glow is due to electric current) [5].

Thus, the idea that often appears in the media that an asteroid may collide with the earth and lead to a cataclysm is another of those lies which humanity's knowledge base is filled with (see Article 1040: Aliens have filled humanity's knowledge base with lies) [6]. The earth is not going to be destroyed by an asteroid, but the earth is at this very moment being destroyed by Planet X drawing energy from the earth's core, and this process is associated to the hatred and evil that the Destroyer, the scourge of the galaxy, Lucifer, telepathically transmits to the minds of mankind (see Article 1132: The Destroyer destroyed the Planet X planets and is now destroying earth) [7].

Figure 21.9. Photograph sent in by Beth of an earth soil creating core emerging from the ground indicating that the earth's core system is losing energy or becoming depleted, which then results in satellite cores moving away from the central core. This is how the earth is getting destroyed and is the same process that led to the Planet X planets breaking into pieces. A cataclysm is in progress on earth as a result.

In conclusion, a rock, which recently appeared on a mound where earth cores are emerging from the ground, due to Planet X, indicates that it came in from space and is part of the Planet X debris field, i.e. it is an asteroid. Asteroids are nothing more than broken pieces of planet and the earth's surface is covered in them. Earth will never be destroyed by a large asteroid impact but the earth is being destroyed by the Planet X absorbing energy from the earth's core system, a process which led to the Planet X planets breaking up.

References:

[1]   Albers, C. (2019). Article 1140: Spherical objects and soil emerging from ground in California due to Planet X.
[2]   Albers, C. (2019). Article 983: Giant Rock in the desert: it came from outer space.
[3]   Albers, C. (2019). Article 1134: Intelligent life all over the Universe.
[4]   Albers, C. (2019). Article 917: Stones fall from the sky in Romania: Planet X debris.
[5]   Albers, C. (2019). Article 779: Meteors are Planet X debris: glow is due to electric current.
[6]   Albers, C. (2019). Article 1040: Aliens have filled humanity's knowledge base with lies.
[7]   Albers, C. (2019). Article 1132: The Destroyer destroyed the Planet X planets and is now destroying earth.

Chapter 22

1192. Fires in California: magma rain due to Planet X in the sky

Fires have broken out over several regions in California at about the same time. The reason why they erupted at the same time is because they are due to a Planet X object coming over the region from space. This particular Planet X core or cores must be in magma creating phase, which means that it or they connect to magma creating cores inside the earth's core system and extract energy from them, which allows the Planet X cores to create the same type of material the earth cores create and also drop the created material, within the gravitational connection region. This is exactly the same process through which these objects create rain in the earth's atmosphere (see Article 695: Planet X creates water: it started at the Flood and Article 1121: Planet X will destroy the earth with magma rain) [1, 2].

Figure 22.1. Fires did not start at one location and spread from there but they started at multiple locations and immediately turned into raging infernos because they were due to material falling from the sky.

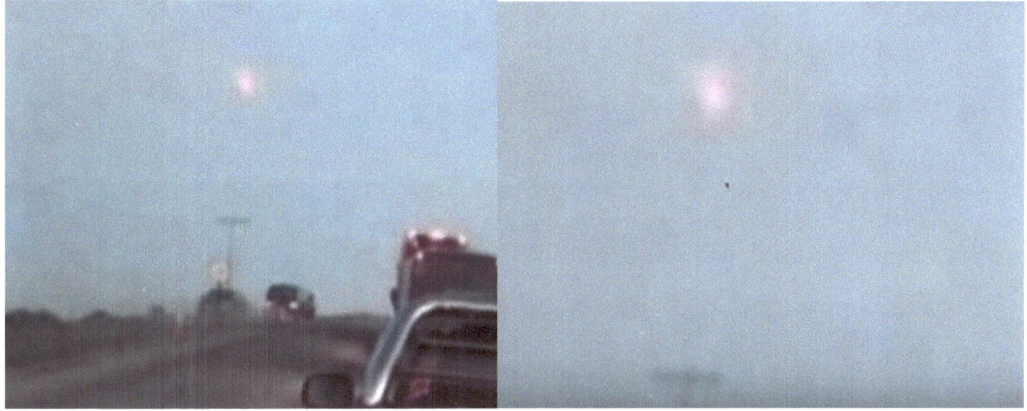

Figure 22.2. Pink object in the sky, one of the several which fell to the surface of the earth and caused fires (**Source:** YouTube Video by MrMBB333 video, 4 October 2019, Something REALLY Strange Collided With Earth - Identity UNKNOWN) (see Article 1177: Magma rain from sky volcanoes due to Planet X) [3].

The material that fell was pink and it started fires where it fell. Meteorites usually look blue or green in the sky and have tails and there are also rocky fragments at the site of the impact. This was magma and as can be seen from the screenshots below magma causes forest fires.

Figure 22.3. Pink magma causes a forest fire in Australia. This material seems to have risen out of the ground as well as fallen from the sky, which is typical of gravitational connections between Planet X cores and earth cores, pink magma creating earth cores, in this case (see Article 757: Planet X causes pink magma to rain down) [4].

Figure 22.4. Yellow, orange and pink flames due to raging fires in California, on October 11th 2019: The terrain is not any different but the flames are different colors and so is the smoke. Why? Because the magma that fell at the different locations was of a different color, most likely because the object or objects in the sky responsible for these fires was or were connected to different magma creating cores at three different connections points, which create magma of different colors. The hottest magma is likely to be the yellow as the yellow flames seem to be more intense.

Thus, the magma creating cores which produce yellow magma are most likely deeper underground, are larger and have a higher gravitational potential than the ones that create orange and pink magma. The coolest magma is most likely the pink as it was observed to rise from the ground in Australia suggesting that these pink magma creating cores are not too far from the surface.

Thus the power companies' desire to switch off power in California just a few days before these fires started due to high winds, according to them, was nothing to do with wind, but in order to avoid being blamed for the fires that are actually due to Planet X coming in from above and now being in magma creating phase (see Article 1186: Power shut down in California to 800 000 customers due to wind?) [5].

Figure 22.5. Screenshot from a video of an object in the sky amongst cloud, which is emitting pink light. The material seems to be semi solid, a consistency typical of magma. The material seems to be forming under a disk shaped cloud that is most likely a cloud envelope for a disk shaped Planet X core and is thus most likely being created by the core within the cloud envelope. The cloud below the pink material is most likely another part of its cloud envelope. This cloud would be analogous to the gaseous material which emerges from the mouth of volcanoes and is referred to as ash. This means that what we are seeing is a volcano in the sky (**Source:** Video by The Two Preachers, 5 October 2019: STRANGE AND UNUSUAL EVENTS IN SAUDI ARABIA | News October 2019 Part 42) [3].

The powers that be have the technology to steer these objects to different locations and therefore their appearance over California and the production of these fires is not by chance, as also the fires created in the Amazon several weeks ago were also not by chance.

Figure 22.6. A fire tornado in Brazil: Tornadoes are created by Planet X, and so a fire tornado can have no other source than Planet X. But this Planet X core or Stellar Core (SC) is not creating just water vapor, it is creating a flammable gaseous substance, which is literally on fire, and it is then using the substance as conduit in the gravitational connection region, in order to absorb energy, from the earth's core system. The core is thus most likely in magma creating phase and is producing incandescent gas as cloud spout material. This seems to therefore be what caused all the fires in the Amazon (see Article 1129: Fire tornado in Brazil: Planet X in magma creation phase) [6].

In conclusion, the erupting of fires at various locations in California are due to magma rain, magma dropping from the surface of Planet X cores, the cores of the destroyed Planet X stars and planets, which are, in this case, connecting gravitationally to earth magma creating cores, which then results in them creating magma and dropping it onto the surface of the earth.

References:

[1]   Albers, C. (2019). Article 695: Planet X creates water: it started at the Flood.
[2]   Albers, C. (2019). Article 1121: Planet X will destroy the earth with magma rain.
[3]   Albers, C. (2019). Article 1177: Magma rain from sky volcanoes due to Planet X.
[4]   Albers, C. (2019). Article 757: Planet X causes pink magma to rain down.
[5]   Albers, C. (2019). Article 1186: Power shut down in California to 800 000 customers due to wind.
[6]   Albers, C. (2019). Article 1129: Fire tornado in Brazil: Planet X in magma creation phase.

Chapter 23

1199. Alarming surface reformation occurring in California due to Planet X and life

Beth has been sending me photographs of the surface reformation event that has been taking place at her property in California, where earth soil creating cores are emerging from the ground and small sinkholes are also constantly appearing, to the point that the ground changes daily. This has allowed me to understand that the earth and indeed the whole universe is alive as the earth's core system grows and procreates in a very similar manner to plant life, on the surface of the planet. This has been detailed in past articles (see Article 1195: Sinkholes: caused by Planet X and show that the universe is alive and Article 1185: Planet X activity in California and the amazing universe we live in) [1, 2]. But in this article that truth sinks in to an even deeper level.

Figure 23.1. Left: Newly emerged earth soil creating core on Beth's property. **Right:** Tiny earth core which was emerging from the ground and Beth picked up. It appears to be about 5 inches in length. It looks like a plant bulb but it is made of soil.

The surface reformation is so changing the soil at this location that some of Beth's plants are suffering. This reformation seems to have started in July, when huge Planet X cores were seen in the sky. They were most likely there before but had not yet approached the surface of the earth closely enough to be seen below the artificial veil that keeps these objects from being seen from the earth's surface and as they approached the surface, the effect on earth's core system seems to have accelerated to an unprecedented rate. The energy absorption process will lead to cores at progressively deeper depths being affected in a similar manner, which will unleash even more dramatic effects, which will be detected from the surface most likely only as earthquakes and this is most likely what is happening in California in regions where this surface activity is not seen. In addition to earthquakes, these effects include much larger sinkholes and possibly the sinking of large sections of surface ground due to underground sinkholes. In regions where surface activity is now seen, effects will also progressively increase as the huge Planet X cores connect to earth cores deeper into the planet, it is likely to lead to magma emerging from the ground and magma falling from the sky, something which seems to already be happening close to Los Angeles (see Article 1192: Fires in California: magma rain due to Planet X in the sky) [3].

Figure 23.2. Tiny earth soil creating core which seems to be made of many different kinds of sandy crystals, including some which are pale green and can be seen to be forming on the end where the core was connected to its parent. However, the layers of soil look a lot like a fungal type of substance.

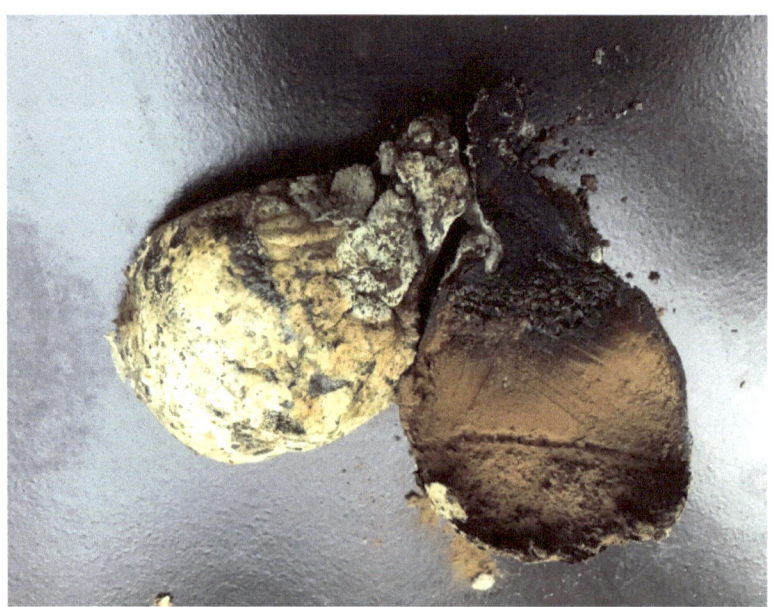

Figure 23.3. When Beth sliced the sliced the tiny core open she could see that it was made of soil: densely packed brown soil.

Figure 23.4. Two new sinkholes, which Beth found, on October 13th 2019: Her property is now full of these small sinkholes which are signs of gravitational connections being made between Planet X cores and the earth's core system, through which they extract energy from the earth's core system. The core responsible for the creation of the soil retracts it or converts it back to energy causing the soil to simply vanish and leaving plant roots untouched.

If the core below the soil had pulled the soil inwards, it would pull the roots down with it but we can see that this is not what happened.

Figure 23.5. Another tiny sinkhole in emerging sandy soil.

Figure 23.6. Beth's butternut squash: it is dried out on the inside, as if the plant had reabsorbed the nutrition within it, most likely because of the changes that occurring to the soil. The squash seems to be sitting on top of newly emerging soil. This shows that Planet X effects will lead to difficulties with the growing of food and thus famine.

Figure 23.7. The outer layer from one of the emerging earth soil creating cores: it looks a lot like the surface you would expect on mushroom and seems to therefore be fungal in nature.

Figure 23.8. Left: The inner part of a soil creating core which emerged some weeks ago, the photograph is from October 8th 2019. **Right:** The same inner part of a soil creating core from October 13th: it seems to be growing an outer layer, which is grey like the layer that the original core had. The material also seems to be growing somewhat like you would expect a fungus to grow on a surface. It too, like the tiny bulb like core has spaces in the connection piece jutting out of the ground.

Figure 23.9. Left: The open spaces in the piece that disappears into the ground which looks similar to the connecting piece on the tiny core. **Right:** The soil creating core, which turned into the jutting piece of hardened soil, seen on the left: It had a grey outer layer or envelope and the soil it thus creates is an extension of its outer layer or envelope. This suggests that the created matter that a planet is made of is core outer envelope material.

The fact that soil creating cores have characteristics similar to fungi suggests that this is the way a planet forms. The core system creates all materials from igneous rock, diamonds, metals, quartz, etc. and toward the surface, the material becomes more and more organic, it becomes fungal in nature, until it turns into edible fungus, on the surface, and then into plant life growing from seeds. Hence, a planet is a living organism.

Figure 23.10. Plant bulbs grow green plants upwards and thus have a wide connection point or opening on top, whilst soil creating cores have the opening on the bottom. But, there are such similarities, between the two, that it can only be concluded that they both work according to the Creator's natural technology: life. The universe is alive.

In conclusion, the intense surface reformation, which is occurring in California, is a sign that Planet X is now extracting energy, from the earth's core system, at an unprecedented rate, which is then causing extreme changes to the planet's surface and most likely deeper underground. The process can only accelerate further and ultimately be cataclysmic for life, living on the surface of the planet. The study of the unfolding cataclysm has however lead to increased understanding of the laws that govern the universe and has led to the conclusion that planets are living organisms, and so are stars and galaxies, as these objects will all contain core systems within them.

References:

[1]   Albers, C. (2019). Article 1195: Sinkholes: caused by Planet X and show that the universe is alive.
[2]   Albers, C. (2019). Article 1185: Planet X activity in California and the amazing universe we live in.
[3]   Albers, C. (2019). Article 1192: Fires in California: magma rain due to Planet X in the sky.

The End for Now!